ENERGY ACCESS, POVERTY, AND D

Ashgate Studies in Environmental Policy and Practice

Series Editor: Adrian McDonald, University of Leeds, UK

Based on the Avebury Studies in Green Research series, this wide-ranging series still covers all aspects of research into environmental change and development. It will now focus primarily on environmental policy, management and implications (such as effects on agriculture, lifestyle, health etc), and includes both innovative theoretical research and international practical case studies.

Also in the series

Rethinking Climate Change Research
Clean Technology, Culture and Communication
Edited by Pernille Almlund, Per Homann Jespersen and Søren Riis
ISBN 978 1 4094 2866 4

A New Agenda for Sustainability
Edited by Kurt Aagaard Nielsen, Bo Elling, Maria Figueroa and Erling Jelsøe
ISBN 978 0 7546 7976 9

At the Margins of Planning
Offshore Wind Farms in the United Kingdom
Stephen A. Jay
ISBN 978 0 7546 7196 1

Contentious Geographies
Environmental Knowledge, Meaning, Scale
Edited by Michael K. Goodman, Maxwell T. Boykoff and Kyle T. Evered
ISBN 978 0 7546 4971 7

Environment and Society
Sustainability, Policy and the Citizen
Stewart Barr
ISBN 978 0 7546 4343 2

Multi-Stakeholder Platforms for Integrated Water Management
Edited by Jeroen Warner
ISBN 978 0 7546 7065 0

Energy Access, Poverty, and Development
The Governance of Small-Scale Renewable Energy in Developing Asia

BENJAMIN K. SOVACOOL
Vermont Law School, USA

IRA MARTINA DRUPADY
*Lee Kuan Yew School of Public Policy,
National University of Singapore, Singapore*

Routledge
Taylor & Francis Group

LONDON AND NEW YORK

First published 2012 by Ashgate Publishing

2 Park Square, Milton Park, Abingdon, Oxon OX14 4RN
711 Third Avenue, New York, NY 10017, USA

Routledge is an imprint of the Taylor & Francis Group, an informa business

First issued in paperback 2016

British Library Cataloguing in Publication Data
Sovacool, Benjamin K.
  Energy access, poverty, and development : the governance of small-scale renewable energy in developing Asia. – (Ashgate studies in environmental policy and practice)
  1. Renewable energy sources—Asia—Case studies. 2. Energy security—Asia—Case studies.
  I. Title II. Series III. Drupady, Ira Martina.
  333.7'94'095–dc23

Library of Congress Cataloging-in-Publication Data
Sovacool, Benjamin K.
  Energy access, poverty, and development : the governance of small-scale renewable energy in developing Asia / by Benjamin Sovacool and Ira Martina Drupady.
     pages cm. -- (Ashgate studies in environmental policy and practice)
  Includes bibliographical references and index.
  ISBN 978-1-4094-4113-7 (hardback : alk. paper)
  1. Renewable energy sources—Asia. 2. Small power production facilities—Government policy—Asia. 3. Energy policy—Asia. 4. Rural development—Asia. I. Drupady, Ira Martina. II. Title.
  TJ807.9.A78S68 2012
  333.79'4095–dc23

2012021243

ISBN 978-1-4094-4113-7 (hbk)
ISBN 978-1-138-26174-7 (pbk)

Contents

List of Abbreviations and Acronyms

$	Denotes United States dollars unless otherwise noted
AC	Alternating current
ADB	Asian Development Bank
AEPC	Alternative Energy Promotion Center
AIDG	Appropriate Infrastructure Development Group
APL	Adaptable Program Loan
ASEAN	Association of Southeast Asian Nations
ASTAE	Asia Sustainable and Alternative Energy Program
AU	Administrative Unit
AusAID	Australian Agency for International Development
BAPPENAS	Badan Perencanaan dan Pembangunan Nasional
BOT	Build-operate-transfer
BPPT	Badan Pengkajian dan Penerapan Teknologi
¢/kWh	Cents per kilowatt-hour
CBO	Community based organization
CBSL	Central Bank of Sri Lanka
CCI	Clinton Climate Initiative
CDM	Clean Development Mechanism
CEB	Ceylon Electricity Board
CFL	Compact Fluorescent Light
CGF	Competitive Grant Facility
CMF	Community mobilization fund
CSP	Concentrated solar power
DC	Direct current
DDC	District development community
DSM	Demand side management
E+Co	Energy Through Enterprise
ECS	Electrical Consumer Society
EdL	Electricité du Laos
EE	Energy Efficiency
EFB	Empty fruit bunch
ENERGIA	International Network on Gender and Sustainable Energy
ESD	Energy Services Delivery Project
ESMAP	Energy Sector Management Assistance Program
EU	European Union
ESCO	Energy service company
GDP	Gross Domestic Product
GEEREF	Global Energy Efficiency and Renewable Energy Fund

GEF   Global Environment Facility
GENI   Global Energy Network Institute
GHG   Greenhouse gas
GIS   Geographic Information System
GIZ   Deutsche Gesellschaft für Internationale Zusammenarbeit (formerly GTZ)
GNESD   Global Network on Energy for Sustainable Development
GS   Grameen Shakti
GTC   Grameen Technology Center
GTZ   Deutsche Gesellschaft für Technische Zusammenarbeit
GVEP   Global Village Energy Partnership
HIV/AIDS   Human Immunodeficiency Virus/Acquired Immunodeficiency Syndrome
IAEA   International Atomic Energy Agency
IAP   Indoor air pollution
IBRD   International Bank for Reconstruction and Development
ICR   Implementation Completion Report
ICS   Improved cookstove
IDA   International Development Association
IEA   International Energy Agency
IED   Innovation Energie Developpement
IFC   International Finance Corporation
IIASS   International Institute for Applied Systems Analysis
IGO   International Governmental Organizations
IMF   International Monetary Funds
INFORSE   International Network for Sustainable Energy
IPP   Independent Power Producer
IRENA   International Renewable Energy Agency
JBIC   Japan Bank for International Cooperation
JICA   Japanese International Cooperation Agency
KeTTHA   Kementerian Tenaga, Teknologi Hijau dan Air
KfW   Kreditanstalt für Wiederaufbau
kg   Kilogram
kW   Kilowatt
kWh   Kilowatt hours
LDC   Least Developed Country
LPG   Liquefied Petroleum Gas
MDGs   Millennium Development Goals
MDSF   Market Development Support Facility
MEM   Ministry of Energy and Mines
MHVE   Microhydro Village Electrification
MNRE   Ministry of New and Renewable Energy
MOFE   Ministry of Fuel and Energy
MW   Megawatt

MWp	Megawatt-peak
NDRC	National Development and Reform Commission
NEA	Nepal Electricity Authority
NGOs	Non-governmental Organizations
NO_x	Nitrous and nitrogen oxides
OECD	Organisation for Economic Co-operation and Development
PAD	Project Appraisal Document
PB	Participating bank
PCI	Participating Credit Institution
PESCO	Provincial Electrification Service Company
PHRD	Policy and Human Resources Development Fund
PIA	Project Implementing Agency
PGI	Provincial Grid Integration
PLN	Perusahaan Listrik Negara
PMO	Project management office
PNG	Papua New Guinea
PO	Participating Organization
POME	Palm oil mill effluent
PPPs	Public private partnerships
PSG	Project Support Group
PV	Photovoltaic
PV-ESCO	Photovoltaic-energy service company
QRF	Quick Response Facility
R&D	Research and development
REAP	Rural Energy Access Project
REB	Rural Electricity Board
REDP	Renewable Energy Development Project (China)
REDP	Rural Energy Development Project (Nepal)
REEEP	Renewable Energy and Energy Efficiency Partnership
REP	Rural Electrification Project
RERED	Renewable Energy Rural Economic Development
RGGVY	Rajiv Gandhi Grameen Vidyutikaran Yojana
SAARC	South Asian Association for Regional Cooperation
S³IDF	Small-Scale Sustainable Infrastructure Fund
SDDC	Song Dian Dao Cun
SDDX	Song Dian Dao Xiang
SEEDS	Sarvodaya Economic Enterprise Development Services
SELCO	Solar Electric Light Company
SHS	Solar home system
SLBDC	Sri Lankan Business Development Center
SLRS	Solar Lighting for Rural Schools
SO_x	Sulfate and sulfur oxides
SPE	Southern Provinces Electrification
SPPA	Small Power Purchase Agreement

SPPT	Small Power Purchase Tariff
SPRE	Southern Provinces Rural Electrification
SREP	Small Renewable Energy Power Program
SWER	Single wire earth return
TI	Technology improvement
TNB	Tenaga National Berhad
TS&L	Teachers Savings and Loan
TSLP	Teachers Solar Lighting Project
TW	Terrawatt
TWh	Terrawatt-hours
UN	United Nations
UNDP	United Nations Development Programme
UNEP	United Nations Environment Programme
UNESCAP	United Nations Economic and Social Commission for Asia and the Pacific
US	United States
USAID	United States Agency for International Development
VCC	Village Development Community
VEC	Village Energy Committee
VEM	Village Electricity Manager
VEP	Village Energy Plan
VESP	Village Energy Security Programme
VDC	Village Development Committee
VOPS	Village Off-grid Promotion and Support
WBCSD	World Business Council on Sustainable Development
WCRE	World Council for Renewable Energies
WTS	Wind turbine system
Wp	Watt-peak
WHO	World Health Organization

List of Figures

List of Tables

Note on Authors

Dr Benjamin K. Sovacool is currently a Visiting Associate Professor at Vermont Law School, where he manages the Energy Security and Justice Program at their Institute for Energy and the Environment. His research interests include the barriers to alternative sources of energy supply such as renewable electricity generators and distributed generation, the politics of large-scale energy infrastructure, designing public policy to improve energy security and access to electricity, and building adaptive capacity and resilience to climate change in least developed Asian countries. He is the author or editor of 12 books and more than 170 peer reviewed academic articles on various aspects of energy and climate change, and he has presented research at more than 80 international conferences and symposia in the past few years. He is a frequent contributor to *Energy Policy*, *Energy & Environment*, *Electricity Journal*, *Energy*, and *Energy for Sustainable Development*.

Ira Martina Drupady is a former Research Associate at the Lee Kuan Yew School of Public Policy in Singapore where she also graduated with a Masters in Public Policy in 2010. Her research work covers energy security, rural electrification, energy development and poverty, and water policy and governance. She is also part of an Asia-Europe policy dialogue exercise to propose reforms to the Institutional Framework for Sustainable Development, recommendations presented at the Rio+20 Earth Summit in 2012.

Foreword

Morgan Bazilian

Special advisor to the Director-General of the United Nation's
program on International Energy and Climate Policy

This book boldly shows us both success and failure in meeting the challenges of expanding access to modern, clean energy services for the poor. That is no small contribution.

Energy powers human progress. From job generation to economic competitiveness, from strengthening security to empowering women, energy is the great integrator: it cuts across all sectors and lies at the heart of all countries' core interests. Now more than ever, the world needs to ensure that the benefits of modern energy are available to all and that energy is provided as cleanly and efficiently as possible. This is a matter of equity, first and foremost, but it is also an issue of urgent practical importance—and this is the impetus for the UN Secretary-General's new *Sustainable Energy for All (SE4All) Initiative.*

This initiative is launched in a time of great economic uncertainty, substantial inequity, high rates of urbanization, and rising youth unemployment. It is also a time where there is emerging consensus on the need to act cohesively towards global issues such as sustainable development. We are not, however, starting from "scratch". New technologies ranging from improved photovoltaic cells, to advanced metering, to electric vehicles and Smart Grids give us a strong foundation from which to move forward. How we capture these opportunities for wealth and job creation, for education and local manufacturing will be the key to unlock any real revolution.

Still, a significant portion of the world's population suffer from a lack of access to affordable basic energy services, such as effective lighting and clean cooking. In the absence of additional dedicated actions, the number of people lacking access to modern energy services will decline only marginally in the coming decades, and will actually increase in some parts of the world, making energy one of the continuing hurdles for development.

The timing of this book is therefore excellent. The UN General Assembly declared 2012 as the *International Year of Sustainable Energy for All* in an effort to catalyze engagement to eliminate energy poverty. But while international attention has increased of late, countries suffering from acute energy poverty have been addressing the issue for decades. Thus, there is myriad action at national and sub-national levels to promote access to modern energy services for the underserved, with varying degrees of success; scaling up such interventions still remains a challenge in most of these areas. Nevertheless, there is an extensive wealth of experience which can, and must, inform the design and implementation of new and more ambitious undertakings. Recent analysis suggests that, despite the progress made, greatly increased efforts (by an order of magnitude) are required to meaningfully address the issue. This book synthesizes many of these lessons in a clear and coherent way.

We must do considerably more than scratch the surface for an issue that deeply impacts all of our lives. The chapters which follow show us exactly how to proceed.

Acknowledgments

The authors are appreciative to the Centre on Asia and Globalization and the Lee Kuan Yew School of Public Policy for some of the financial assistance needed to conduct the research interviews, field research, and travel for this project. The authors are also extremely grateful to the National University of Singapore for Faculty Start-up Grant 09-273 as well as the MacArthur Foundation for Asia Security Initiative Grant 08-92777-000-GSS, which have supported elements of the work reported here. Any opinions, findings, and conclusions or recommendations expressed in this material are those of the authors and do not necessarily reflect the views of the Centre on Asia and Globalization, Lee Kuan Yew School of Public Policy, National University of Singapore, or MacArthur Foundation.

We are also immensely grateful to Anthony L. D'Agostino from Columbia University in the United States, Malavika Jain Bambawale from Bosch in Singapore, Olivia Gippner from the Freie Universität Berlin in Germany, Saroj Dhakal from the National University of Singapore in Singapore, Martin Stavenhagen from Physikalisch-Technische Bundesanstalt in Germany, and Debajit Palit from The Energy and Resources Institute in India for helping collect some of the primary data for our case studies.

Furthermore, portions of the book draw from arguments and data published previously in "Bridging the Gaps in Global Energy Governance," *Global Governance* 17(1) (January–March, 2011), pp. 57–74; "The Socio-Technical Barriers to Solar Home Systems (SHS) in Papua New Guinea: 'Choosing Pigs, Prostitutes, and Poker Chips over Panels,'" *Energy Policy* 39(3) (March, 2011), pp. 1532–1542; "Gers Gone Wired: Lessons from the Renewable Energy and Rural Electricity Access Project (REAP) in Mongolia," *Energy for Sustainable Development* 15(1) (March, 2011), pp. 32–40; "Conceptualizing Urban Household Energy Use: Climbing the 'Energy Services Ladder,'" *Energy Policy* 39(3) (March, 2011), pp. 1659–1668; "Realizing Rural Electrification in Southeast Asia: Lessons from Laos," *Energy for Sustainable Development* 15(1) (March, 2011), pp. 41–48; "Halting Hydro: A Review of the Socio-Technical Barriers to Hydroelectric Power Plants in Nepal," *Energy* 36(5) (May, 2011), pp. 3468–3476; "An International Comparison of Four Polycentric Approaches to Climate and Energy Governance," *Energy Policy* 39(6) (June, 2011), pp. 3832–3844; "Innovation in the Malaysian Waste-to-Energy Sector: Applications with Global Potential," *Electricity Journal* 24(5) (June, 2011), pp. 29–41; "Summoning Earth and Fire: The Energy Development Implications of Grameen Shakti in Bangladesh," *Energy* 36(7) (July, 2011), pp. 4445–4459; "Security of Energy Services and Uses within Urban Households," *Current Opinion in Environmental Sustainability* 3(4) (September, 2011), pp. 218–224; "Bending Bamboo:

Restructuring Rural Electrification Strategy in Sarawak, Malaysia," *Energy for Sustainable Development* 15(3) (September, 2011), pp. 240–253; "Electrification in the Mountain Kingdom: The Implications of the Nepal Power Development Project (NPDP)," *Energy for Sustainable Development* 15(3) (September, 2011), pp. 254–265; "Examining the Small Renewable Energy Power (SREP) Program in Malaysia," *Energy Policy* 39(11) (November, 2011), pp. 7244–7256; "And Then What Happened?: A Retrospective Appraisal of China's Renewable Energy Development Project (REDP)," *Renewable Energy* 36(11) (November, 2011), pp. 3154–3165; "Improving Access to Modern Energy Services: Insights from Case Studies," *Electricity Journal* 25(1) (January, 2012), pp. 93–114; "What Moves and Works: Broadening the Consideration of Energy Poverty," *Energy Policy* 42 (March, 2012), pp. 715–719; "New Partnerships and Business Models for Facilitating Energy Access," *Energy Policy* 47 (June, 2012), pp. 48–55; "A Comparative Analysis of Solar Home System Programs in China, Laos, Mongolia and Papua New Guinea," *Progress in Development Studies* (in press, 2012); "Unsold Solar: A Post-Mortem of Papua New Guinea's Teachers Solar Lighting Project," Journal of Energy & Development 35 (1/2) (in press, 2012); "Peeling the Energy Pickle: Expert Perceptions on Overcoming Nepal's Electricity Crisis," *South Asia: Journal of South Asian Studies* (in press, 2012);

Chapter 1
Introduction

Introduction

Energy poverty—lack of access to electricity and dependence on solid biomass fuels for cooking and heating—remains an enduring global problem. Approximately 1.4 billion people still live without electricity, and an additional 2.7 billion people depend entirely on wood, charcoal, and dung for their domestic energy needs.[1] Lack of access to modern energy not only limits opportunities for income generation and blunts efforts to escape poverty, it also severely impacts women and children and contributes to global deforestation and climate change. The search for energy fuels and services is therefore an arduous, daily grind for billions of people around the world, most of them in Asia, where serious repercussions of energy poverty assume different forms depending on geographical terrain, population size, and climatic variations.

However, small-scale renewable energy technologies—solar home systems, residential wind turbines, biogas digesters and gasifiers, microhydro dams, and improved cookstoves—offer these households and communities the ability to tackle extreme poverty, enhance gender equality and education, reduce hunger, provide safe drinking water, improve health, and ensure environmental sustainability. Innovative collaborations and programs involving governments as well as businesses, nonprofit organizations, banks, and community based cooperatives have emerged in recent years to expand access to these technologies and the energy services they offer. All over Asia, these burgeoning partnerships have come in different forms: some focus on improving technological performance, others on providing low-cost loans, still others leasing out systems according to a "fee-for-service" model. But regardless of their approach, such technologies and the programs that support them can drastically improve living standards for some of the poorest communities in the region.

Based on extensive field research, our book showcases how these small-scale renewable energy technologies are helping Asia respond to a daunting set of energy security challenges. The book offers a compendium of the most interesting renewable energy case studies over the last ten years from one of the most diverse regions in the world. This book examines ten different case studies of the developing and scaling up of renewable energy technologies, six of them successes, four of them failures, in Bangladesh, China, Laos, Mongolia, Nepal, Sri Lanka, India, Indonesia, Malaysia, and Papua New Guinea.

---

1 International Energy Agency, United Nations Development Programme, United Nations Industrial Development Organization, 2010. *Energy Poverty: How to Make Modern Energy Access Universal?* Paris: OECD.

What Makes This Book Special

As the United Nations International Year of Sustainable Energy kicks off this year in 2012, four things set this book apart from other research works supporting efforts to expand energy access, improve efficiency and increase the use of renewable energy worldwide.

First and foremost, the book focuses intensely on the topic of household energy security in developing countries, mostly in rural areas. Roughly one-third of the entire population of developing countries consumes less than the equivalent of $1 per day in goods and services; one-fifth of them have no access to modern healthcare services; one-third lack safe drinking water; and two-thirds lack access to sanitation.[2] These regions therefore face entirely distinct energy security threats than the ones confronting industrialized economies such as the United States or those comprising the European Union.

Although energy poverty is a global problem of epidemic proportions, it is frequently neglected in energy planning discussions and academic publications. Already more than a decade ago, Daniel Kammen and Michael R. Dove wrote that advanced and modern technologies related to electricity and motorized transport (such as "nuclear reactors" and "electric vehicles") were highly favored topics of energy research yet "mundane" technologies such as cookstoves, biogas units, heating and cooling systems, and other less "state-of-the-art" topics were minimally investigated, even though these technologies affected the greatest number of people and had the most substantial impact on the environment in everyday life.[3] Ten years later, Fatih Birol, the Chief Economist for the International Energy Agency (IEA), argued that "unfortunately, the energy-economics community has given far less attention to the challenge of energy poverty among the world's poorest people."[4] And most recently, a series of content analyses of the top energy journals noted that only three percent of authors came from least developed countries and less than eight percent of papers addressed topics related to energy poverty and energy development.[5]

This lack of focus is alarming, for without access to electricity, many households across the developing world must rely on candles, biomass, and kerosene lamps

---

2   United Nations Development Programme 1997. *Energy After Rio: Prospects and Challenges.* Geneva: United Nations.

3   Kammen, D.M. and Dove, M.R. 1997. The Virtues of Mundane Science. *Environment* 39(6) (July/August, 1997), 10–41.

4   Birol, F. 2007. Energy Economics: A Place for Energy Poverty on the Agenda? *The Energy Journal* 28(3), 1–6.

5   See D'Agostino, A.L. et al. 2011. What's the State of Energy Studies Research?: A Content Analysis of Three Leading Journals from 1999–2008. *Energy* 36(1) (January, 2011), 508–519; and Sovacool, B.K. et al. 2011. What About Social Science and Interdisciplinarity? A 10-year Content Analysis of Energy Policy. in *Tackling Long-Term Global Energy Problems: The Contribution of Social Sciences*, edited by D.L. Goldblatt et al. New York: Springer.

that burn fuel for lighting. One in three people in the world obtain light from such fuels and pay $40 billion per year, or 20 percent of global lighting costs, to do so. In spite of bearing these costs, they only receive 0.1 percent of the world's lighting energy services.[6] In the realm of cooking, "conventional" or "traditional" stoves emit a great deal of smoke into the home (akin to "living constantly inside a giant cigarette"), which can cause acute respiratory illnesses among inhabitants that regularly lead to death.

Second, the book offers a neutral, critical evaluation of notable energy access projects harnessing renewable energy sources. Neither of the authors is associated with any of the projects studied nor the institutions involved with them. We are thus able to offer dispassionate, objective analysis of the strengths and weaknesses evident in different national approaches and programs. As part of this critical analysis, our book discusses six case studies of success, commonly referred to as "best practices," alongside four case studies of failure, or "worst practices." Our inquiry comes at a time when the need for a rigorous, independent and comprehensive evaluation of renewable energy access efforts is becoming more apparent. One recent study found, for example, that no less than 140 countries (including 50 least developed countries) have some type of national energy access target or program, figures reflected in Table 1.1.[7] Yet most of these programs lack independent monitoring and evaluation, so it remains difficult to share credible information about best technologies and practices.[8] Our book fills this gap.

Third, our book is comparative, looking at ten countries and cases, and socio-technical, meaning we examine not only technologies but also the influence of things such as social behavior, governance regimes, regulations, and price signals. All case studies uncover the increasingly conventional wisdom of polycentric

Table 1.1   Developing countries with energy access targets

	Developing Countries	Least Developed Countries
Electricity	68	25
Modern Fuels	17	8
Improved Cookstoves	22	4
Mechanical Power	5	0
Total	140	50

---

6   Mills, E. 2006. *Alternatives to Fuel-based Lighting in Rural Areas of Developing Countries*. Berkeley: Lawrence Berkeley National Laboratory.

7   Legros, G. et al. 2009. *The Energy Access Situation in Developing Countries: A Review Focusing on the Least Developed Countries and Sub-Saharan Africa*. New York: World Health Organization and United Nations Development Program, 29.

8   Bazilian, M. 2010. More Heat and Light. *Energy Policy* 38, 5409–5412.

approaches, where desired elements such as equity, inclusivity, accountability, and adaptability are better achieved in seamless power sharing between overlapping and multiple scales of governance, mechanism and actors.[9]

Put another way, our research sits at the nexus between technology, politics, development, and energy security.[10] We identify not only the programmatic factors that often result in the success or failure of individual case studies, but also the complex agendas of international and bilateral energy and development agencies, manufacturers, research planners, politicians, and community leaders. The narrative that results is more complex, though we believe accurate, than assessments and reports that commonly assess only one or two of these variables in isolation.

Lastly, our books adds to the body of knowledge on renewable energy development through primary data collection that arises from original extensive research interviews with more than 400 energy experts, representing roughly 200 institutions, over a period of four years. These interviews have been supplemented by an exhaustive review of contemporary scientific and technical literature on energy governance, access, and security issues, as well as by 90 site visits and consultations with 781 community members. Our book therefore allows for a rich and meaningful analysis that addresses the breach between perceptions of stakeholders and facts on the ground experienced by the energy poor themselves.

The remainder of this introductory chapter begins by investigating the concepts of energy poverty, the energy ladder, and energy equity. It then details the most up-to-date statistics on rural energy use and energy poverty in Asia before summarizing the research methods utilized in the book. The final part of the introduction previews the case studies and chapters to come.

Conceptualizing Energy Poverty

As there is no simple definition of poverty, conceptualizing "energy poverty" is a somewhat arduous process. Recent work, including the United Nations Development Programme's Human Development Report, has noted that poverty is not a static or fixed state, but instead a multi-dimensional concept encompassing caloric intake, life expectancy, housing quality, literacy, access to energy, and a variety of other factors.[11] Such poverty is frequently expressed from an income perspective: to be "poor" is to earn less than $2 per day when adjusted for the purchasing power parity of countries. Under this definition, a shocking 40 percent

---

9   Sovacool, B.K. 2011. An International Comparison of Four Polycentric Approaches to Climate and Energy Governance. *Energy Policy* 39, 3832–3844.

10   Martinot, E. 2001. Renewable Energy Investment by the World Bank. *Energy Policy* 29, 689–699.

11   United Nations Development Programme 2010, *Human Development Report 2010* New York: UNDP.

of the global population is poor.[12] Sticking with the UNDP's multidimensional notion of poverty, factors such as health, education, and living conditions can be just as important as sources of employment or wages. Within this list of non-income dimensions, two energy indicators are found: electricity (having no electricity constitutes poverty) and cooking fuels (relying on wood, charcoal, and/ or dung for cooking constitutes poverty). Such a conception of energy poverty has been confirmed by the IEA and other multilateral organizations which state that energy poverty is comprised of lack of access to electricity and reliance on traditional biomass fuels for cooking.[13]

Thus, the UNDP explicitly defines energy poverty as the "inability to cook with modern cooking fuels and the lack of a bare minimum of electric lighting to read or for other household and productive activities at sunset."[14] The Asian Development Bank takes a slightly broader approach to defining energy poverty, and tells us that it is "the absence of sufficient choice in accessing adequate, affordable, reliable, high-quality, safe and environmentally benign energy services to support economic and human development."[15]

Several ways of measuring such energy poverty exist. One method is to track the minimum amount of physical or animate energy needed for basic needs such as cooking and lighting, often including the minimum amount of food needed to lead to a healthy, nutritious life. Another is to look at the poorest people in a given country, say households in the lowest income quintile, and to then detail the types and amounts of energy they use. Yet another is measuring how much income is spent on energy services; typically a family that spends more than ten to 15 percent of their earnings on energy services per month or year is considered "energy poor" or classified as in "fuel poverty."[16]

The IEA's "Energy Development Index" is composed of four indicators which each "capture a specific aspect of potential energy poverty":

---

12   D'Agostino, A.L. 2010. Energy Insecurity for ASEAN's BoP: The Un-electrified 160 Million, presentation to the *ESI-MINDEF Workshop*, Singapore, October 5.

13   International Energy Agency, United Nations Development Programme, United Nations Industrial Development Organization 2010, *Energy Poverty: How to Make Modern Energy Access Universal?* Paris: OECD; Jones, R. 2010. Energy Poverty: How to Make Modern Energy Access Universal? *Special Early Excerpt of the World Energy Outlook 2010 for the UN General Assembly on the Millennium Development Goals* Paris: International Energy Agency/OECD.

14   Gaye, A. 2007. Access to Energy and Human Development. *Human Development Report 2007/2008*. United Nations Development Programme Human Development Report Office Occasional Paper, 4.

15   Masud, J., Sharan, D., and Lohani, B.N. 2007. *Energy for All: Addressing the Energy, Environment, and Poverty Nexus in Asia*. Manila: Asian Development Bank, April, 47.

16   Dutta, S. 2011. *Sustainable Energy Development in the Asia Pacific: A Discussion Note*. Bangkok: UNESCAP.

- per capita commercial energy consumption, which serves as an indicator of the overall economic development of a country;
- per capita electricity consumption in the residential sector, which serves as an indicator of the reliability of, and consumer's ability to pay for, electricity services;
- share of modern fuels in total residential sector energy use, which serves as an indicator of the level of access to clean cooking facilities; and
- share of population with access to electricity.[17]

The 2012 "Global Energy Assessment" from the International Institute for Applied Systems Analysis also has an entire cluster (roughly one-quarter of their lengthy report) dedicated to energy poverty under the title "Realizing Energy for Sustainable Development."[18]

The most common concept illustrating energy poverty involves the "energy ladder." One study defines the energy ladder as "the percentage of population among the spectrum running from simple biomass fuels (dung, crop residues, wood, charcoal) to fossil fuels (kerosene, natural gas, and coal direct use) to electricity."[19] The idea implies that the primary types of energy used in rural areas or developing countries can be arranged on a "ladder" with the "simplest" or most "traditional" fuels and sources, such as animal power, candles, and wood, at the bottom with the more "advanced" or "modern" fuels such as electricity or refined gasoline at the top. The ladder is often described in terms of efficiencies, with the more efficient fuels or sources occupying higher rungs. For example, kerosene is three to five times more efficient than wood for cooking, and liquefied petroleum gas is five to ten times more efficient than crop residues and dung.[20] Table 1.2 depicts the energy ladder as discussed in a variety of academic studies.[21]

---

17   International Energy Agency 2011. *The Energy Development Index*. Paris: IEA. Available at: http://www.iea.org/weo/development_index.asp.

18   IIASA 2011. *Global Energy Assessment*. Laxenberg: IIASA. Available at: http://www.iiasa.ac.at/Research/ENE/GEA/.

19   John P. Holdren and Kirk R. Smith. "Energy, the Environment, and Health," in Tord Kjellstrom, David Streets, and Xiadong Wang (Eds.) *World Energy Assessment: Energy and the Challenge of Sustainability* (New York: United Nations Development Programme, 2000), pp. 61–110.

20   Barnes, D.F. and Floor, W.M. 1996. Rural Energy in Developing Countries: A Challenge for Economic Development. *Annual Review of Energy and Environment* 21, 497–530.

21   See International Energy Agency, United Nations Development Programme, United Nations Industrial Development Organization 2010; Jones 2010; Legros, G. et al. 2009. *The Energy Access Situation in Developing Countries: A Review Focusing on the Least Developed Countries and Sub-Saharan Africa*. New York: World Health Organization and United Nations Development Programme; Cook, C. et al. 2005. *Assessing the Impact of Transport and Energy Infrastructure on Poverty Reduction*. Manila: Asian Development Bank, 249; International Energy Agency 2004. *World Energy Outlook 2004*. Paris: OECD; and Barnes and Floor 1996.

Table 1.2   The energy ladder

Sector	Energy Service	Developing Countries			Developed Countries
		Low-income Households	*Middle-income Households*	*High-income Households*	
Household	Cooking	Wood (including wood chips, straw, shrubs, grasses, and bark), charcoal, agricultural residues, and dung	Wood, residues, dung, kerosene, and biogas	Wood, kerosene, biogas, liquefied petroleum gas, natural gas, electricity, and coal	Electricity and natural gas
	Lighting	Candles and kerosene (sometimes none)	Candles, kerosene, paraffin, and gasoline	Kerosene, electricity, and gasoline	Electricity
	Space Heating	Wood, residues, and dung (often none)	Wood, residues, and dung	Wood, residues, dung, coal, and electricity	Oil, natural gas, or electricity
	Other Appliances	None	Electricity, batteries, and storage cells	Electricity	Electricity
Agriculture	Tilling or Plowing	Hand	Animal	Animal, gasoline, and diesel (tractors and small power tillers)	Gasoline and diesel
	Irrigation	Hand	Animal	Diesel and electricity	Electricity
	Post-harvest Processing	Hand	Animal	Diesel and electricity	Electricity
Industry	Milling and Mechanical	Hand	Hand and animal	Hand, animal, diesel, and electricity	Electricity
	Process Heat	Wood and residues	Coal, charcoal, wood, and residues	Coal, charcoal, kerosene, wood, residues, and electricity	Coke, napthene, and electricity
Primary Technologies		Cookstoves, three-stone fires, and lanterns	Improved cookstoves, biogas systems, solar lanterns, and incandescent and compact fluorescent light bulbs	Improved cookstoves, biogas systems, liquefied petroleum gas, gas and electric stoves, compact fluorescent light bulbs, and light emitting diodes	

The Advisory Group on Energy and Climate Change, an intergovernmental body composed of representatives from businesses, the United Nations, and research institutes, divide energy access into incremental categories. First comes basic human needs met with both electricity consumption of 50 to 100 kWh per person per year and 50 to 100 kg of oil equivalent or modern fuel per person per year (or the ownership of an improved cookstove). Second comes productive uses such as access to mechanical energy for agriculture or irrigation, commercial energy, or liquid transport fuels. Consumption of electricity here rises to 500 to 1,000 kWh per year plus 150 kg of oil equivalent. Third comes modern needs which include the use of domestic appliances, cooling and space heating, hot and cold water, and private transportation which in aggregate result in the consumption of about 2,000 kWh of electricity per year and 250 to 450 kg of oil equivalent. Table 1.3 illustrates this sequential ordering quite clearly.[22]

Table 1.3   Energy services and access levels

Level	Electricity use	kWh per person per year	Solid fuel use	Transport	Kilograms of oil equivalent per person per year
Basic human needs	Lighting, health, education, and communication	50 to 100	Cooking and heating	Walking or bicycling	50 to 100
Productive uses	Agriculture, water pumping for irrigation, fertilizer, mechanized tilling, and processing	500 to 1,000	Minimal	Mass transit, motorcycle, or scooter	150
Modern society needs	Domestic appliances, cooling, and heating	2,000	Minimal	Private transport	250 to 450

Regardless of the way it is depicted, the energy ladder suggests a rising gap between how the rich and poor consume energy—with implications on equity and affordability. The so-called "richest" tend to consume much more energy—as much as 21 *times* more—than the lowest quintiles, the so-called "poor"; which means access to energy, or lack of it, can both reflect and worsen social inequality.[23] Rural households tend to be poorer and consume much less energy than urban households, and in rural areas, fuelwood, the most common source of energy, is usually harvested in unsustainable ways with severe impacts on forest health and the health of those using it, something discussed below. Another study looking at Asia noted that the poor typically pay more for energy needs yet receive poorer quality energy services due to inefficient and more polluting technologies with higher upfront costs.[24]

The State of Asian Energy Poverty

Notwithstanding these complexities, organizations such as the IEA, the World Health Organization, and various United Nations organizations have done a remarkable job compiling statistics on energy poverty.

According to the most recent data available, as of 2009, 1.4 billion people lack access to electricity, 85 percent of them in rural areas, and almost 2.7 billion people remain reliant on woody biomass fuels for cooking, numbers broken down in Table 1.4. An additional one billion people have access only to unreliable or intermittent electricity networks.[25] Put another way, the poorest three-quarters of the global population still only use ten percent of global energy.[26]

In Asia specifically, energy access oscillates noticeably. China alone accounts for about 30 percent of the electricity generated for the entire region, and six countries—Australia, China, India, Japan, Korea, and Russia—account for 87 percent of generated electricity. When broken down into per capita figures, houses in New Zealand or Australia consume 100 times more electricity than those in Bangladesh and Myanmar.[27] As Table 1.4 also reveals, 55 percent of those without access to electricity globally as well as 72.3 percent of those dependent on traditional fuels globally reside in Asia.

Other recent statistics confirm these trends. About one billion people live below $1.25 per day in the Asia Pacific, 70 percent of all Asian poor are women, and 900 million workers in Asia earn less than $2 per day, leading one recent study to proclaim that "the state of human deprivation compels us to consider a paradigm shift to universal energy access and a minimal standard for quality of life. Energy

---

23   United Nations Development Programme 2009. *Contribution of Energy Services to the Millennium Development Goals and to Poverty Alleviation in Latin America and the Caribbean*. Santiago, Chile: United Nations.

24   Masud et al. 2007.

25   United Nations Development Programme 2010, 7.

26   Bazilian, M. et al. 2010. More Heat and Light. *Energy Policy* 38, 5409–5412.

27   United Nations Economic and Social Commission for Asia and the Pacific 2010. *Lighting up Lives: Pro-Poor Public Private Partnerships*. Bangkok, Thailand: UNESCAP.

Table 1.4   Number of people without access to electricity and dependent on traditional fuels

	Number of People Lacking Access to Electricity (millions)	Number of People Relying on the Traditional Use of Biomass for Cooking (millions)
Africa	587	657
Sub-Saharan Africa	585	653
Asia	799	1,937
China	8	423
India	404	855
Other Asia	387	659
South America	31	85
World	1,417	2,679

Sources: International Energy Agency, United Nations Development Programme, United Nations Industrial Development Organization 2010.

security policies must be pro-poor."[28] An independent UNDP study concurred, noting that the urban poor typically have some access to electricity but its quality is substandard, service unreliable and intermittent, and connections informal. The rural poor often go without modern energy services entirely and when they do have access, it tends to be from inefficient standalone diesel systems, poorly run micro-grids that are expensive and susceptible to failure, or patchy connections to the national grid.[29]

All the while, the environmental costs of energy production and use in Asia continue to rise. China is now the world's largest emitter of greenhouse gases, responsible for about one-quarter of the world's total in 2008, and India has more than doubled its carbon emissions from 1990 to 2008.[30] Since 1990, emissions from coal alone have increased 185 percent in China, and 141 percent in India.[31]

---

28   Krairiksh, N. 2011. The Social Dimensions of Energy Security in the Asia-Pacific, presentation to the United Nations Economic and Social Commission for Asia and the Pacific (UNESCAP) Expert Group Meeting on *"Sustainable Energy Development in Asia and the Pacific,"* United Nations Convention Center, Bangkok, Thailand, September 27–29.

29   United Nations Development Programme 2011. *Energy for People-Centered Sustainable Development.* New York: UNDP.

30   United Nations 2011. Carbon dioxide emissions (CO2), thousands metric tons of CO2 (CDIAC), Millennium Development Goals Indicators. *The Office United Nations Cite for the MDG Indicators.* Available at: http://mdgs.un.org/unsd/mdg/SeriesDetail. aspx?srid=749&crid= [accessed: July 7, 2011].

31   International Energy Agency 2010. CO2 emissions from fuel combustion: highlights, 47, 49.

This voracious demand for energy in Asia is predominately driven by two factors, or "twin culprits." The first is "consumption-led" demand as classes of Asian people achieve increases in luxury and standards of living, bringing with them more energy-intensive lifestyles that revolve around automobiles, air conditioning, and disposable goods. The second is "industrial-led" demand related to economic growth, a structural shift from non-mechanized forms of manufacturing and production to more energy intensive ones, especially for commodities such as iron and steel, cement and glass, paper and pulp, basic chemicals, and nonferrous metals.[32] Looking to the future, analysts expect that developing countries in Asia, driven by China and India, will raise their share in global energy consumption from 24 percent in 2005 to 35 percent in 2030, while this share for OECD countries will decline from 52 percent to 41 percent, respectively.[33]

Perhaps surprisingly, the growth of electricity demand in some Southeast Asian countries has actually outpaced growth in China and India. Over the period 1985 to 2005, per capita electric power consumption increased much faster in Southeast Asian countries than the world average. This consumption level rose by a factor of 1.6 for the world average but by a factor of 5.0 in China, 8.1 in Vietnam, 6.3 in Indonesia, 4.8 in Thailand, 3.7 in Malaysia, 2.4 in Singapore, and 1.7 in the Philippines (See Table 1.5). Table 1.5 also indicates that China and most Southeast Asian countries (except for Singapore and Malaysia) were still far below the world average in terms of electricity use.

Table 1.5 **Level of per capita consumption of electricity for the world average, China and ASEAN-6 countries (USD in 2000 = 100)**

	1985	1995	2005
World	12.0	16.3	19.7
China	2.6	5.6	13.0
ASEAN-6			
Vietnam	0.5	1.1	4.2
Indonesia	0.6	2.0	3.7
Thailand	3.0	9.4	14.5
Malaysia	6.4	14.8	23.9
Singapore	25.2	44.4	61.2
Philippines	2.6	3.0	4.3

32 Rosen, D.H. and Houser, T. 2007. *China Energy: A Guide to the Perplexed.* Washington, DC: Center for Strategic and International Studies.

33 Energy Information Administration 2008. *International Energy Outlook 2008,* DOE/EIA-0484(2008), Table 1.

Thus, the current situation of energy access is somewhat contradictory: more people have access to electricity as a percentage of the global population, but that access comes with debilitating social, economic, and environmental costs; there are also more people without energy access in absolute terms. Moreover, future trends could make things worse. By 2030, the number of people relying on traditional biomass will rise from 2.7 billion today to 2.8 billion.[34] By that same year, according to the newest projections, one-third of the global population will still be dependent on biomass for cooking, 1.3 billion will lack access to reliable electricity networks, and two-thirds of those people will reside in just two regions: Africa and Asia.[35] The World Bank estimates that poor households around the world spend about $20 billion per year on traditional fuels for cooking and lighting.[36] In India alone, the annual rural energy services market is worth an estimated $2 billion.[37] To achieve the UN's ambitious target of universal energy access by 2030, $36 billion would need to be invested in energy poverty and electrification efforts *each year* between now and then.[38]

Renewable Energy to the Rescue

Our research in Asia shows that a collection of small-scale renewable energy systems can literally and seriously complement efforts to provide billions of people in the region with reliable energy services and lift them out of poverty. Three of these services are the most urgent at fulfilling basic needs and increasing the earning potential of rural households: lighting, heating and cooking, and mechanical power.

Lighting

One in three people in the world obtain light from "traditional" fuels, yet dependence on traditional fuel-based technologies for lighting has dire consequences for the world's poorest people.

First, traditional "fuel-based" lighting technologies are more expensive. After 50,000 hours of use, kerosene lamps cost $1,251 to operate, while incandescent lamps cost $175, compact fluorescent lamps cost $75, and white light emitting

---

34   Jones 2010.

35   Bazilian et al. 2010.

36   World Bank 2011. Pro-Poor Energy Access Technical Assistance Programs. Available at: http://www.esmap.org/esmap/PEA-TAP.

37   World Resources Institute 2010. *Power to the People: Investing in Clean Energy for the Base of the Pyramid in India.* Washington, DC: WRI, 2.

38   International Energy Agency, United Nations Development Programme, United Nations Industrial Development Organization 2010.

diodes $20 to operate.[39] Moreover, rural households spend as much as one-quarter of their household budgets on fuel for illumination without even taking into account losses in productivity and other indirect expenses.[40]

Second, using traditional fuel-based technologies has severe health implications. The World Bank estimates that 780 million women and children inhale particulate-laden kerosene fumes while performing daily tasks in their homes. Kerosene fumes contain nitrogen oxides, sulfur oxides, and volatile organic compounds, which cause eye, nose, throat, and lung infections, respiratory problems, and cancer to those that inhale them. Additionally, fuel-based lighting contributes to severe burns and accidental fires.

Third, traditional fuel-based lighting has negative environmental impacts and contributes to climate change. A single kerosene lantern, for example, emits 40 times as much carbon dioxide as an incandescent lamp and 180 times as much as a CFL.[41] Fuel-based lighting in the developing world is a source of "244 million tons of carbon dioxide emissions into the atmosphere each year, or 58 percent of the carbon dioxide emissions from residential electric lighting."[42]

Fourth, fuel-based lighting provides a poor quality of light. A kerosene lamp offers only two to four lumens compared to a 60-watt bulb, which provides 900 lumens of light. A paraffin wax candle has a lighting intensity of one lumen and efficiency in terms of lumen per watt of 0.01, whereas a 15-watt CFL has an intensity of 600 lumens and an efficiency of 40 lumens per watt.[43] The low level of light produced by kerosene and candles is undesirable for performing ordinary tasks best done during the nighttime and creates a barrier to education for children who cannot study after dark.

Thankfully, communities and households can utilize biogas digesters, solar photovoltaic panels, small wind energy systems, and microhydro dams to provide light without connecting to a fossil-fueled electric-grid or relying on liquid fuels. The electricity from these distributed systems can be used to power incandescent lamps, CFLs, and WLEDs. While households continue to use incandescent lamps widely, WLED and CFL technologies use less energy, provide better light, and have significantly longer lamp life.[44]

39   Pode 2010.

40   Adkins, E. et al. 2010. Off-Grid Energy Services for the Poor: Introducing LED Lighting in the Millennium Villages Project in Malawi. *Energy Policy* 38, 1087–1097.

41   Mills 2006.

42   Pode 2010.

43   Jones 2010.

44   Pode, R. 2010. Solution to Enhance the Acceptability of Solar-powered LED Lighting Technology. *Renewable and Sustainable Energy Reviews.* 14, 1096–1103.

Heating and Cooking

Electricity accounts for roughly 17 percent of global final energy demand, yet low temperature heat accounts for about 44 percent. Globally, this means that people use more energy for heating—typically by burning woody biomass—than for any other purpose.[45]

The majority of households in the developing world—about three out of every four—rely on traditional stoves for their cooking and heating needs. Traditional stoves range from three-stone open fires to brick and mortar models and ones with chimneys.[46] These stoves emit a significant amount of smoke into the home, which can cause acute respiratory illnesses among inhabitants. Indoor air pollution is especially problematic for women and children who spend the most time cooking. Some may be exposed to as much as 200 times the recommended level of small particulates, the most dangerous type of air pollution.[47]

Reliance on traditional stoves can also cause significant environmental problems. Because these stoves are highly inefficient (as much as 90 percent of their energy content is wasted), they require a significant amount of fuel—almost two tons of biomass per family per year. These consumption patterns strain local timber resources and can cause "wood fuel crises" when wood is harvested faster than it is grown.[48] The problem of fuel shortages and ecosystem losses is compounded where industrial logging and agriculture put an additional—and often more intense—pressure on forest resources.[49] Furthermore, recent studies suggest that the burning of biomass also exacerbates climate change. Households in developing countries burn about 730 tons of biomass annually, which translates into more than one billion tons of carbon dioxide.[50] As an illustration, Japan, currently the fifth largest carbon dioxide emitter in the world, emitted 1.2 billion tons of carbon dioxide in 2008.[51]

Improved cookstoves offer a promising alternative. Though "improved stove" is a broad term, it generally includes stoves that improve energy efficiency, remove indoor air pollution, and reduce the "drudgery" of fuel gathering. ICS alleviate

---

45   Olz, S., Sims, R., and Kirchner, N. 2007. Contributions of Renewables to Energy Security. *International Energy Agency Information Paper*. Paris: OECD.

46   World Bank 2011a. *Household Cookstoves, Environment, Health & Climate Change: A New Look at an Old Problem*. Washington, D.C.: World Bank.

47   World Health Organization 2005. *Air Quality Guidelines: Global Update 2005*. Copenhagen: WHO.

48   Crewe, E., Sundar, S. and Young, P. 2010. *Building a Better Stove: The Sri Lanka Experience*. Colombo.

49   Barnes D.F. et al. 1994. What Makes People Cook With Improved Biomass Stoves? *World Bank Technical Paper #242: Energy Series*.

50   World Health Organization 2005.

51   The World Bank 2012 *Data*. Available at: <http://data.worldbank.org/indicator/ EN.ATM.CO2E.KT?order=wbapi_data_value_2008+wbapi_data_value+wbapi_data_ value-last&sort=asc [accessed: January 23, 2012].

many of the public health and environmental problems that stem from widespread use of primitive, traditional stoves.[52] One assessment compared the economic costs of investing in new cookstoves—including the expense of fuel, program costs, and capital costs for technology—to their corresponding benefits, such as reduced health care expenses, productivity gains, time savings, and improvement of the environment.[53] It studied these costs and benefits in 11 developing countries from 2005 to 2015, and found that if these countries switched to 50 percent coverage of ICS, the results would be:

- substantial reduction of acute lower respiratory infections, chronic obstructive pulmonary diseases, and lung cancer;
- increased numbers of illness free days and deaths avoided;
- time savings from reduced needs to survey fuel areas and collect fuel, as well as quicker cooking times;
- avoided deforestation; and
- avoided carbon dioxide and methane emissions.

The study concluded that the benefit-to-cost ratio of investing in ICS was extremely high, with an investment of $650 million producing $105 billion in benefits per year. Similarly, the IEA noted that switching to LPG stoves at the global scale would cost about $13.6 billion but would produce annual benefits exceeding $91 billion.[54]

Mechanical Power

Mechanical power increases the efficiency and effectiveness of productive activities supporting sustainable development, as well as physical processes fundamental to meeting basic human needs. Mechanical services have great potential to tremendously reduce time spent on fuelwood gathering, to improve air quality in homes, and to raise household and community incomes.

Many of these benefits have a great impact on the lives of women and children who traditionally spend time gathering fuel and water and cooking in the home, or performing tasks related to milling, grinding, and husking. In India, women in rural households spent almost three hours cooking daily meals, an additional hour or two more processing food to make it ready for cooking, and, when needed, fuel collection, which tends to consume two more hours of time. In Nepal, without

---

52 Bazilian, M. et al. 2011. Partnerships for Access to Modern Cooking Fuels and Technologies. *Current Opinion in Environmental Sustainability* 3, 1–6.

53 Hutton, G., Rehfuess, E., and Tediosi, F. 2008. Evaluation of the Costs and Benefits of Household Energy and Health Interventions, presentation to the *Clean Cooking Fuels & Technologies Workshop*, June 16–17, Istanbul, Turkey.

54 International Energy Agency 2006. *World Energy Outlook 2006*. Paris: OECD.

mechanical energy, women have to get up well before dawn, process daily agricultural requirements, and then cook meals, meaning access to electricity does not really save them time.

Whereas electricity produced by microhydro units tends to be used by one-third to one-half of all villagers, agricultural processing units tend to be used by 92 to 97 percent of villagers. This means that microhydro units oriented toward mechanical processing can create massive social benefits, saving 30 to 110 hours per person per month for various processing requirements such as milling, hulling, and expelling.

Mechanical power enables activities such as pumping, transporting, and lifting water, irrigating fields, processing crops, small-scale manufacturing, and natural resource extraction. As one recent study concluded:

> Experiences show that mechanical power helps alleviate drudgery, increase work rate and substantially reduce the level of human strength needed to achieve an outcome, thus increasing efficiency and output productivity, producing a wider range of improved products, and saving time and production costs ... In this regard, financing of mechanical power is often one of the most cost effective ways to support poor people. [55]

These examples strongly suggest that some of the most fundamental services required for reducing poverty and promoting human development involve mechanical energy and increasing the productivity of human labor.

Case Selection and Research Methods

How, then, ought these benefits of renewable energy technologies be captured?

To better comprehend the dynamics of successful and unsuccessful renewable energy access programs, we began by selecting cases to review. The idea was to select countries with similar economies, projects of a similar type, and technology of a similar scale. We decided that we wanted case studies in the geographic Asia-Pacific and also for countries generally classified as lower-middle income or below—that is, they have per capita gross domestic product of about $4,000 per year and below, though Malaysia is an exception.

We started by exploring the academic literature on these types of projects as well as reports and case studies on energy poverty published by a variety of energy institutions summarized in Table 1.6. This initial review produced a staggering amount of data, with more than 1,100 projects and programs related to renewable energy systems published in the last two decades.

---

55   Bates, L. et al. 2009. Expanding Energy Access in Developing Countries: The Role of Mechanical Power. *Practical Action Report*. Washington, D.C.: UNDP.

Table 1.6 Global renewable energy and energy poverty actors

Institution	Acronym	Central Location	Primary Function	Description
United Nations System	UN	New York, USA; Vienna, Austria; Geneva, Switzerland	Building international peace and security as well as promoting social progress, better living standards, and human rights	Both core UN bodies (e.g. UNEP, UNDP, and UNESCAP) as well as loosely affiliated specialized agencies (e.g. Food and Agricultural Organization; the International Atomic Energy Agency) working on various energy issues. The newly formed umbrella UN Energy is intended to coordinate their efforts
Global Environment Facility	GEF	Washington DC, USA	As the world's largest public environment fund, the GEF sponsors environmental projects, through grants to developing countries, for biodiversity, climate change, international waters, deforestation, and biodiversity loss. GEF was made independent of the World Bank in 1994	GEF is the entrusted financier for projects for the United Nations Framework Convention on Climate Change as well as several other international conventions relating to energy. It has so far allocated almost $9 billion in funds including $40 million as part of a Least Developed Countries Fund for Climate Change and the Special Climate Change Fund
International Energy Agency	IEA	Paris, France	To establish a reporting system on oil prices and create an emergency oil-sharing system, and to serve as a key information source on energy	The IEA has been relatively successful at coordinating national action among oil consuming countries, although membership excludes such key oil consumers as China and India. It is the primary producer of global energy statistics and is moving to address broader energy and climate topics
International Renewable Energy Agency	IRENA	Abu Dhabi, UAE	Charged with promoting renewable energy among its 142 member countries	Although only about two years old, IRENA has already begun developing a knowledge base of best practices for renewable energy promotion, providing policy advice, facilitating technology transfer and financing, and stimulating research on all aspects of renewable energy

continued ...

Institution	Acronym	Central location	Primary function	Description
World Business Council on Sustainable Development	WBCSD	Geneva, Switzerland	Also founded at the 1992 Rio Earth Summit, WBCSD is a global association of some 200 companies and 55 partner and regional organizations dealing with business and sustainable development that sees its primary function as advocating for businesses and influencing policy	Aims to create a platform for companies to explore sustainable development best practices, share knowledge, and advocate business positions. Also manages a variety of business sponsored projects including energy efficiency in buildings, water, cement, electricity supply, forest products, mining and minerals, and tires
Global Village Energy Partnership	GVEP	London, UK	Seeks to reduce poverty through accelerated access to modern energy services through its 2,000-plus members which include a mix of private companies, national governments, development agencies, multilateral financial institutions, and universities	Committed to forming partnerships from the bottom up at the community and municipal level to increase energy access and also build capacity to adapt to climate change
Clinton Climate Initiative	CCI	New York, USA	Part of the William J. Clinton Foundation, CCI manages an extensive program to undertake building retrofits, improve outdoor lighting, reduce waste, measure GHG emissions, encourage non-motorized transport, and promote "climate positive" communities in major cities, conducts research on carbon capture and storage and concentrating solar power, and works with Cambodia, Guyana, Kenya, Indonesia, and Tanzania to prevent deforestation	Brings stakeholders from industry (such as energy service contractors and the manufacturers of energy efficient equipment), the public sector (municipal and city governments), and finance (banks and lending agencies) to conduct climate-related projects in 40 metropolitan areas; Forestry project has also teamed up with university research institutes and government agencies

Name	Abbreviation	Location	Description	Achievement
Energy Through Enterprise	E+Co	Bloomfield, New Jersey, USA	Focuses on clean energy innovation by partnering multilateral financial institutions with NGOs and the private sector through eight international offices to implement projects in 20 developing countries	Provides debt and equity to support the expansion of energy services to rural populations around the world through the use of entrepreneurs
Global Energy Efficiency and Renewable Energy Fund	GEEREF	European Investment Bank, Luxembourg	Created by the European Commission to promote public-private partnerships in clean energy through private equity funds to small- and medium-sized enterprises in emerging economies	Has so far leveraged or disbursed about $200 million in more than 20 projects in the developing world
Small-Scale Sustainable Infrastructure Development Fund	S³IDF	Cambridge, Massachusetts, USA	Promotes a Social Merchant Bank approach to help local entrepreneurs create micro-enterprises that provide infrastructure services to the poor	Has so far built a portfolio of almost 200 small investments and associated enterprises in India with an additional 100 projects in the pipeline
World Energy Council	WEC	London, UK	Charged with promoting sustainable energy and energy access through research and analysis, energy projections, and recommendations in 93 countries	Produces publications, hosts conferences, and arranges meetings covering all major energy sources, including electrification and off-grid sources, as well as a World Energy Conference once every three years
International Fund for Agricultural Development	IFAD	Rome, Italy	Combats rural hunger and poverty in developing countries through low-interest loans and direct assistance	Works with the rural poor, governments, donors, and NGOs to improve rural access to biogas and SHS units, and to reduce drudgery and "lighten the load" for rural women
Solar Electric Light Fund	SELF	Washington, DC, USA	Created to empower people in developing countries to escape poverty harnessing energy from the sun	Has established more than a dozen self-sustaining solar energy projects in eleven countries spread across Asia, Africa, and South America
Acumen Fund	AF	New York, USA	Formed to reduce poverty by investing in social enterprises and "breakthrough" ideas in the health, water, housing, energy, and agriculture sectors	Approves about $6 million per year in social enterprise funds for microhydro, solar, biogas, biomass, and lighting projects in India, Pakistan, and East Africa

Since this was simply too much information, we relied on a seven-phase selection process to reduce the number of case studies to a manageable number.

First, to be included in our study, a program had to involve the direct provision or supply of energy services through renewable energy to rural or poor communities and areas. This meant, in practice, that programs distributing other end-use devices, such as light bulbs or mobile phones, were ineligible. This reduced our case study pool to 944.

Second, a case study had to be a fully implemented program in operation for at least four years. The intent here was to exclude pilot and demonstration projects, and also projects so short lived it would be difficult to draw general lessons from them. This reduced our pool to 332.

Third, we excluded case studies that did not promote the five core technologies described above (solar home systems, wind turbines, biogas digesters or gasifiers, microhydro units, and cookstoves). This criterion meant that only small-scale technology, below 10 MW of installed capacity, was included. Excluded were renewable energy projects at the centralized, electric-utility scale. This lowered our pool to 290.

Fourth, we wanted programs of a moderate size. So we excluded case studies that had budgets of less than $50,000, which distributed energy services or technologies to less than 750 homes or customers, and/or those that installed less than 100 kW of total capacity. This dropped the number to 117.

Fifth, a case study had to be recent, either currently in operation or completed in the past ten years when we started the project in 2008. This meant we excluded all projects ending before the calendar year 2000, lowering our pool to 55.

Sixth, sufficient data had to exist on the case study in question. This was admittedly subjective, but generally we excluded cases with less than five published sources of credible information. This did create a bias in favor of cases from the World Bank, the GEF, and the UN, as they are extensively documented. This reduced our pool to 24.

Seventh and lastly, the case in question had to either be a clear-cut example of success or failure. By success, we mean it accomplished its goals, at or below cost, and before or on schedule. By failure, we mean it did not accomplish its goals, was above cost, and/or completely behind schedule. The literature on case study selection calls these "extreme" cases, for they study those at the outermost end of success or failure. This left us with the 10 case studies summarized in Table 1.7.

The idea here was to select extreme cases that could possibly have wider lessons beyond an individual country. Certainly, this method is only one of the many ways in which we could have arrived at the final selection case studies. We could have cast a wider net that included more countries, projects, or approaches and used different selection criteria. However, the ten case studies that we have included ensures that we are able to look at renewable energy access and development throughout a mix of geographic regions, cultures, electricity networks, procurement mechanisms, regulations, standards, technologies, income groups, and project durations. At the same time, we are confident that we have developed a robust selection framework that can be applied to incorporate more case studies in the future.

With our cases selected, we then relied on what academics call an inductive, narrative case study approach based on data collected from research interviews and field research. The primary reason we chose such qualitative methods is that few peer reviewed studies existed on our ten cases and little academic literature looks explicitly at energy use in developing Asian economies, requiring us to collect even basic data about rural electrification, energy production, and household energy use. A semi-structured research interview format enabled us to ask experts involved with each case study a set of standard inquiries but then allowed the conversation to build and deviate to explore new directions and areas. We also relied on qualitative methods because many of the variables of interest to us such as the ongoing energy policy challenges facing each country or the factors explaining the success or failure of its renewable energy programs, are difficult to measure and cannot be described purely with numerical analysis. In addition, our case studies involved site visits and discussions with rural community leaders and some of these participants were illiterate, making textual collection of data impossible.

In aggregate, we conducted 441 of these research interviews and meetings with 200 institutions over the course of four years, research trips summarized in Table 1.8.[56] In each case we had simultaneous real-time translation into local languages and dialects. We relied on a purposive sampling strategy, meaning experts with extensive knowledge of each case were chosen to participate, and also a critical stakeholder analysis framework that required us to include respondents from government, civil society, business, academia, and local communities, as well as people in favor, and opposed to, each project. We made sure to include participants from:

- *Government agencies* such as the Nepal Ministry of Energy, Indonesian Ministry of Finance, Indian Ministry of New and Renewable Energy, Chinese Ministry of Science and Technology, or Sri Lanka Sustainable Energy Authority;
- *Intergovernmental organizations* such as the South Asian Association for Regional Cooperation, the Global Environment Facility, and the United Nations Development Programme;
- *International civil society organizations* or *think tanks*, including Conservation International, Friends of the Earth, Transparency International, and the Stockholm Environmental Institute;
- *Local civil society organizations* or *think tanks*, including Grameen Shakti, Yayasan Pelangi Indonesia, and Pragati Pratishthan;

---

56   We did not undertake all of this research alone, and had some extraordinary assistance from our friends and colleagues. Readers should peruse the acknowledgments section of the book for specific details.

Table 1.7 Overview of our ten case studies

Type	Country	Case study	Primary partners	Primary beneficiaries	Technology	Dates	Cost ($m)	Accomplishments
Success	Bangladesh	Grameen Shakti	Grameen Shakti, Grameen Technology Centers, World Bank, Infrastructure Development Corporation Limited, and Government of Bangladesh	Rural communities and women	Solar home systems, biogas, and improved cookstoves	1996 to present	$100 (annual)	Installation of 500,000 solar home systems, 132,000 cookstoves, and 13,300 biogas plants among 3.1 million beneficiaries.
Success	China	Renewable Energy Development Program	World Bank, GEF, National Development and Reform Commission, and local solar manufacturers	Manufacturers and nomadic herders	Solar home systems	2002 to 2007	$316	Distributed more than 400,000 units in 5 years
Success	Laos	Rural Electrification Project	Electricité du Laos, Ministry of Energy and Mines, World Bank, GEF, and PESCOs	Rural communities and PESCOs	Small hydro and solar home systems	2006 to 2009	$13.75	Electrified 65,000 previously off-grid homes and disbursed more than 17,000 solar home systems
Success	Mongolia	Rural Energy Access Project	World Bank, National Renewable Energy Center, Ministry of Fuel and Energy, Ministry of Finance, and Ministry of Environment and Resources	Nomadic herders, *soum* centers, and local solar companies	Solar home systems and wind turbines	2007 to 2011	$23	Distributed 41,800 solar home systems, hundreds of wind turbines, facilitated the rehabilitation of 15 mini grids in soum centers, installed 11 renewable-diesel hybrid systems
Success	Nepal	Rural Energy Development Program	World Bank, Government of Nepal, UNDP, Nepal Alternative Energy Promotion Center, District Development Communities, Village Development Communities, and Microhydro Functional Groups	Rural communities	Microhydro	2004 to 2011	$5.5 (original proposal)	Distributed 250 units benefitting 50,000 households in less than 10 years

	Country	Project	Funders/Partners	Technology	Years	Cost ($M)	Outcome
Success	Sri Lanka	Energy Services Delivery Project	World Bank, GEF, Ceylon Electricity Board, and national banks	Solar home systems and microhydro	1997 to 2002	$55.3	Installed 21,000 solar home systems and 350 kilowatts of village hydro capacity in rural Sri Lanka, in addition to 31 megawatts of grid-connected mini-hydro capacity
Failure	India	Village Energy Security Project	Ministry of New and Renewable Energy, World Bank, and Village Energy Committees	Biomass gasifiers, biogas systems, and improved cook stoves	2004 to 2011	$8.6	Aimed to install 61 biogas projects, only 21 of 50 projects still functioning by 2009
Failure	Indonesia	Solar Home System project	World Bank, GEF, Ministry of Energy and Mineral Resources, Perusahaan Listrik Negara, and local banks	Solar home systems	1997 to 2003	$118.1	Aimed to install 200,000 SHS across one million users; only 8,054 SHS units ever installed reaching 35,000 villagers
Failure	Malaysia	Small Renewable Energy Power Program	Ministry of Energy, Green Technology, and Water, Tenaga Nasional Berhad, and Sabah Electricity Sendirian Berhad	Solar home systems, microhydro, biogas, and waste incineration	2001 to 2010	$220 (unconf.)	Tried to install 500 MW of small-scale renewable energy technology by 2005, but ended up achieving only 12 MW. Target altered to 350 MW by 2010, but only 61.7 MW were built by project close
Failure	Papua New Guinea	Teachers Solar Lighting Project	World Bank, GEF, Department of Education, Papua New Guinea Sustainable Energy Limited, and Teacher's Savings & Loan	Solar home systems	2003 to 2010	$2.9	Attempted to install 2,500 solar home systems and jumpstart a local market, ended up installing only 1 single unit

- *Electricity suppliers* including the Nepal Electricity Authority, Tenaga Nasional Berhad in Malaysia, Ceylon Electricity Board in Sri Lanka, and Papua New Guinea Power Limited;
- *Manufacturers, industry groups*, and *commercial retailers* such as Alstrom Hydro, Barefoot Power Systems, Sime Darby, Siemens, and Sunlabob;
- *Financiers* and *bilateral development donors* including Deutsche Gesellschaft für Technische Zusammenarbeit, United States Agency for International Development, the Asian Development Bank, and the World Bank Group; and
- *Universities* and *research institutes* including the International Center for Integrated Mountain Development, University of Dhaka, University of Papua New Guinea, and the Chinese Academy of Sciences.

This list is not exhaustive, and readers are invited to see Appendix 1 for all institutions interviewed.

Table 1.8   Overview of research interviews

Country	Case Study	Number of Interviews	Number of Institutions	Dates Visited
Bangladesh	Grameen Shakti	48	19	Jun. 2009 to Oct. 2010
China	Renewable Energy Development Program	30	17	May 2010 to Jun. 2010
Laos	Rural Electrification Project	16	11	Mar. 2010
Mongolia	Rural Energy Access Project	22	10	June.2010
Nepal	Rural Energy Development Program	57	24	Aug. 2010 to Nov. 2010
Sri Lanka	Energy Services Delivery Project	56	28	Feb.2011
India	Village Energy Security Project	51	17	Sep.2008 to Jun. 2009
Indonesia	Solar Home System project	36	22	Jun. 2011
Malaysia	Small Renewable Energy Power Program	89	38	Mar. 2010 to Feb. 2011
Papua New Guinea	Teachers Solar Lighting Project	36	14	Feb. 2010 to Apr. 2010
Total		441	200	Sep. 2008 to Jun. 2011

For each case study, we asked participants to (a) Identify the benefits of the program at hand, (b) summarize some of the key barriers to implementation it had to confront, and (c) discuss general lessons that the case study offers energy policy and development practitioners. Due to Institutional Review Board guidelines at the National University of Singapore, as well as the request of some participants, we present such data in our book as anonymous, though information from the interviews was often recorded and always carefully coded.

To ensure a degree of triangulation and reliability, and to better understand our case studies, we augmented our research interviews with direct observation and site visits to ninety renewable energy facilities in our ten countries over the course of March 2009 to June 2011 (see Table 1.9), some of them shown in Figures 1.1 to 1.3. These included a variety of different sources, systems, sizes, and capacities, including some grid-connected facilities to gain a comparative perspective, as well as laboratories, testing centers, factories, and assembly lines. The site visits enabled us to discuss our cases with actual renewable energy operators, managers, and manufacturers. They also served as a useful vehicle to arrange additional research interviews. Moreover, during our research trips we spoke with almost 800 community members, village leaders, and households in aggregate including local political representatives, interactions summarized in Table 1.10.

We supplemented our interviews, site visits, and community consultations with a review of reports and peer-reviewed articles relating to energy policy in each country (though there were not many of these).

By "case studies" we endeavored to provide what methodological theorists Alexander George and Andrew Bennett call a "detailed examination of an aspect of a historical episode to develop or test historical explanations that may be generalizable to other events."[57] Rather than utilizing laboratory samples or statistical analysis to examine variables, case study methods involve an in-depth, longitudinal assessment of a single instance or group of instances: a case or cases.[58] Put another way, the case study method is an investigation of a contemporary phenomenon within its real-life context to explore causation in order to find underlying principles.[59]

---

57   George, A.L. and Bennett, A. 2004. *Case Studies and Theory Development in the Social Sciences*. Cambridge, MA: Harvard University Press, 5.

58   See Flyvbjerg, B. 2006. Five Misunderstandings About Case Study Research. *Qualitative Inquiry* 12(2), 219–245; and Flyvbjerg, B. 2001. *Making Social Science Matter: Why Social Inquiry Fails and How It Can Succeed Again*. Cambridge, MA: Cambridge University Press.

59   Yin, R.K. 2003. *Case Study Research: Design and Methods*. London: Sage.

Table 1.9   Summary of renewable energy site visits

Name	Type of facility	Capacity	Owner/Operator	Location	Date visited
Various (fifty separate community systems)*	Biogas, SHS	-	Various communities	The states of Assam, Chhattisgarh, Gujarat, Madhya Pradesh, Maharashtra, Orissa, and West Bengal	Mar. 2009 to Nov. 2009
Nam Theun 2	Hydroelectric	1,070 MW	Nam Theun 2 Power Corporation	Khammouane, Laos	Mar. 2010
Westlink Solar Enterprises	SHS distribution	-	Westlink Solar	Port Moresby, Papua New Guinea	Mar. 2010
Sikikoge Primary School	SHS	160 Wp	Sikikoge Primary School	Goroka, Papua New Guinea	Mar. 2010
Madang Elementary School	SHS	220 Wp	Madang Elementary School	Madang, Papua New Guinea	Mar. 2010
Monmat Solar Energy	SHS distribution	-	Monmat Solar Energy Ltd	Ulaanbaatar, Mongolia	Jun. 2010
HK Mart Solar	SHS distribution	-	HK Mart	Ulaanbaatar, Mongolia	Jun. 2010
Green Eco-Energy Park Project	SHS, mini-hydroelectric, and wind	110 kW	Daesung Group	Nalayh, Mongolia	Jun. 2010
Photovoltaic and Wind Power Systems Quality Test Center	SHS and wind turbines	-	Chinese Academy of Sciences	Beijing, China	Jun. 2010
Beijing Jike New Energy Technology Development Company	SHS manufacturing and assembly	-	Beijing Jike	Beijing, China	Jun. 2010
Gesang Solar Energy Company	SHS manufacturing and assembly	-	Gesang Solar	Xining, China	Jun. 2010
Qinghai Tianpu Solar Energy Company	SHS manufacturing and assembly	-	Tianpu Solar	Xining, China	Jun. 2010
Xining Moonlight Solar Science and Technology Co.	SHS manufacturing and assembly	-	Xining Moonlight Solar	Xining, China	Jun. 2010

Project	Type	Capacity	Developer	Location	Date
Xining New Energy Development Co. (NIDA)	SHS manufacturing and assembly	-	NIDA	Xining, China	Jun. 2010
Nanhui Wind Farm	Wind	21 MW	Shanghai Electric	Shanghai, China	Jun. 2010
Batang Ai Hydroelectric Station	Hydroelectric	108 MW	Sarawak Energy Bhd.	Batang Ai, Sarawak, Malaysia	Jul. 2010
Bakun Hydroelectric Project	Hydroelectric	2,400 MW	Ministry of Finance/Sarawak Hydro Bhd.	Bakun, Sarawak, Malaysia	Jul. 2010
Murum Hydroelectric Project	Hydroelectric	944 MW	Sarawak Energy Bhd.	Murum, Sarawak, Malaysia	Jul. 2010
Kg. Mudung Abun Microhydro Plant	Microhydro	25 kW	Mudung Abun Community	Denang, Sarawak, Malaysia	Jul. 2010
Long Lawen Microhydro Plant	Microhydro	10 kW	Long Lawen Community	Long Lawen, Sarawak, Malaysia	Jul. 2010
Lubok Antu Palm Oil Mill	Palm Oil	1 MW	Salcra Sdn. Bhd.	Sri Aman, Sarawak, Malaysia	Jul. 2010
Grameen Technology Center - Singair	Improved cookstoves (manufacturing)	-	Grameen Shakti	Singair, Bangladesh	Oct. 2010
Grameen Technology Center - Mawna	Biogas (manufacturing and installation)	-	Grameen Shakti	Mawna, Bangladesh	Oct. 2010
Malekhola Microhydro Village Electrification Project	Microhydro	26 kW	Malekhola Village Development Committee	Malekhola, Nepal	Nov. 2010
Daunme Khola Microhydro Village Electrification Project	Microhydro	12 kW	Daunme Khola Village Development Committee	Daunme Khola, Nepal	Nov. 2010
Bom Khola Microhydro Village Electrification Project	Microhydro	100 kW	Bom Khola Village Development Committee	Bom Khola, Nepal	November 2010
Sungai Kerling Minihydro Plant	Minihydro	2 MW	Renewable Power Sdn Bhd.	Kerling, Selangor, Malaysia	Jan. 2011
Langkawi Cable Car Solar-Diesel Hybrid	Solar photovoltaics/ Diesel	109.5 kW	Langkawi Development Authority and Tenaga Nasional Bhd.	Pulau Langkawi, Kedah, Malaysia	Jan. 2011

continued ...

Name	Type of facility	Capacity	Owner/Operator	Location	Date visited
Hybrid Integrated Renewable Energy System	Solar/Wind/Diesel	400 kW	State Government of Terengganu and Tenaga Nasional Bhd.	Pulau Perhentian, Terengganu, Malaysia	Jan. 2011
TDM Palm Oil Estate	Palm oil	1.0 MW	TDM Plantation Sdn. Bhd.	Dungun, Terengganu, Malaysia	Jan. 2011
Kajang Waste-to-Energy Plant	Waste incineration	8.9 MW	Core Competences Sdn. Bhd., Recycle Energy Sdn. Bhd.	Semenyih, Selangor, Malaysia	Jan. 2011
Bukit Tagar Sanitary Landfill	Landfill gas capture	1 MW	Kub-Berjaya Enviro Sdn. Bhd.	Bukit Tagar, Selangor, Malaysia	Jan. 2011
Bell Palm Oil Mill	Palm Oil Mill Effluent, methane capture and Empty Fruit Bunch incineration	1.7 MW Gas Capture, 10 MW Combustion (under construction)	Bell Eco Power Sdn. Bhd. and Bell Palm Industries Sdn. Bhd.	Batu Pahat, Johor, Malaysia	Feb. 2011
Meddawatte Village Hydro System	Microhydro	12 kW	Meddawatte Village Electricity Board	Sabaragmuwa Province, Sri Lanka	Feb. 2011
Watawala Hydroelectric Power Plant	Hydroelectric	2.7 MW	Mark Marine Services (Pvt) Limited	Central Province, Sri Lanka	Feb. 2011
Carolina Estate Minihydro Project	Hydroelectric	2.5 MW	Caroline Tea Estate	Central Province, Sri Lanka	Feb. 2011
LOLC Solar Power Plant	Solar (recycled SHS)	48 kW	Lanka Orix Leasing Company Ltd	Colombo, Sri Lanka	Feb. 2011
Hambantota Wind Farm	Wind	3 MW	Ceylon Electricity Board	Hambantota, Sri Lanka	Feb. 2011
Trimba Solar Demonstration Facility	SHS (solar modules, street lamps, batteries)	-	PT. Trimbasolar	Jakarta, Indonesia	Jun. 2011

* These site visits were conducted by colleagues at The Energy and Resources Institute (TERI) in India.

Table 1.10   Summary of community discussions and focus groups

Case Study	Name/Location	Technology	No. of discussions (approximate)	Date visited
India	Assam	Biogas and SHS	14	Mar. 2009
India	Chhattisgarh	Biogas	6	Apr. 2009
India	Gujarat	Biogas	2	May 2009
India	Madhya Pradesh	Biogas and SHS	10	Jun. 2009
India	Maharashtra	Biogas	5	Jul. 2009
India	Orissa	Biogas	9	Sep. 2009
India	West Bengal	Biogas and SHS	4	Nov. 2009
Laos	Mai Village, Keo Oudom District	SHS	100	Mar. 2010
Laos	Nongsa, Nazay Thong District	Microhydro	30	Mar. 2010
Papua New Guinea	Akameku	SHS	10	Mar. 2010
Papua New Guinea	Asaroka	SHS and wind	15	Mar. 2010
Papua New Guinea	Lufa	SHS	25	Mar. 2010
Papua New Guinea	Kundiawa	SHS	25	Mar. 2010
Papua New Guinea	Okifa	SHS	20	Mar. 2010
Papua New Guinea	Simbu	SHS	15	Mar. 2010
Papua New Guinea	Talidig	SHS	20	Mar. 2010
China	Qinghai Province	SHS	40	Jun. 2010
Mongolia	Terelj	SHS	10	Jun. 2010
Mongolia	Nalaikh	SHS	15	Jun. 2010
Mongolia	Tsonjinboldog	SHS and wind	5	Jun. 2010
Malaysia	Selangor	SHS	5	Jul. 2010
Malaysia	Asap	SHS	40	Jul. 2010
Malaysia	Bakun	SHS	15	Jul. 2010
Malaysia	Upper Bakun	SHS, ICS	25	Jul. 2010
Malaysia	Danang	Microhydro	5	Jul. 2010
Malaysia	Murum	SHS	50	Jul. 2010
Malaysia	Lubok Antu	Microhydro	5	Jul. 2010
Bangladesh	Singair	ICS and biogas	20	Oct. 2010
Bangladesh	Manikguni	ICS and SHS	15	Oct. 2010
Bangladesh	Mawna	Biogas and SHS	15	Oct. 2010
Nepal	Changdol	Microhydro	10	Nov. 2010
Nepal	Chongba	Microhydro	15	Nov. 2010
Nepal	Kavre	Microhydro	10	Nov. 2010
Nepal	Lukla	Microhydro	12	Nov. 2010

continued ...

Case study	Name/Location	Technology	No. of discussions (approximate)	Date visited
Nepal	Dhading	Microhydro	14	Nov. 2010
Nepal	Puchetar	Microhydro	18	Nov. 2010
Sri Lanka	Meddawatte Village	Microhydro	10	Feb. 2011
Sri Lanka	Indigolla Village	SHS	20	Feb. 2011
Sri Lanka	Dagama Village	SHS	15	Feb. 2011
Sri Lanka	Ponnilawa Village	SHS	25	Feb. 2011
Indonesia	Jangari Village, West Java	SHS	10	Jun. 2011
Indonesia	Lake Cirata, West Java	SHS	25	Jun. 2011
Indonesia	Serdang Village, Lampung	SHS	30	Jun. 2011
Total	43 Communities		789 participants	

We present the information collected from our case studies in a narrative format. We rely primarily on a narrative presentation of data because narratives, or storylines, are an elemental part of understanding human behavior. Narratives, or in our case "narrative analysis," documents the "raw" world as it is experienced by its subjects, and it is most appropriate for capturing what actual energy users or consumers believe.[60] Such an inductive, narrative, case study approach has been used widely in the fields of public policy and energy policy to gain insight into the dynamics of energy programs, the different perceptions of stakeholders, and consumer acceptance (or rejection) of specific energy technologies.

The biggest benefit of such an approach is that narrative case studies have the "irreducible quality" of being "thick," that is, full of rich detail, or "hard to summarize".[61] The biggest drawback is that the resulting story, or narrative, can be at times complicated and unstructured, though we believe it is also a more accurate reflection of reality.

We also present in this study a number of photographs and images related to energy use collected during our fieldwork. This is because we believe that such images act as important "physical evidence" that can develop a more precise understanding of the topic being studied.[62] Photographs can uncover a sort of "visual ethnography" that reveals the meaning behind events in ways that words

60   Czarniawska, B. 2004. *Introducing Qualitative Methods: Narratives in Social Science Research*. London: Sage.
61   Lijphart, A. 1971. Comparative Politics and the Comparative Method. *American Political Science Review* 65 (1971), 682–693.
62   Yin 2003.

Figure 1.1   The research team interviewing operators of the Watawala Hydroelectric Power Plant, Sri Lanka

Figure 1.2   The research team speaking with a nomadic herder using a SHS near Qinghai, China

Figure 1.3   The research team visiting a biogas digester near Singair, Bangladesh

cannot.[63] Because we experience our world through both words and images, the exclusion of "visual elements" artificially limits the narrative.[64] As methodological experts Prosser and Schwartz have written:

> [Photographs] can show characteristic attributes of people, objects, and events that often elude even the most skilled wordsmiths. Through our use of photographs we can discover and demonstrate relationships that may be subtle or easily overlooked. We can communicate the feeling or suggest the emotion imparted by activities, environments, and interactions. And we can provide a degree of tangible detail, a sense of being there and a way of knowing that may not readily translate into other symbolic modes of communication.[65]

Though the practice of including images in research is more common in the disciplines of advertising, business, marketing, anthropology, sociology, communication studies, and rhetoric, we believe those presented in our book help augment our presentation of textual data.

63   Pink, S. 2007. *Doing Visual Ethnography: Images, Media, and Representation in Research*. London: Sage.

64   Birdsell D.S. and Groarke L. 1996. Toward a Theory of Visual Argument. *Argumentation and Advocacy* 33(1), 1–10.

65   Prosser J. and Schwartz D. 1998. Photographs Within the Sociological Research Process, in *Image-Based Research: A Sourcebook for Qualitative Researchers* edited by J. Prossner. New York: Routledge, 115–130.

What's Next

The chapters to come investigate the technologies involved in fighting energy poverty, our ten case studies, and two final chapters offering comparative lessons and conclusions. Each case study chapter is structured similarly, beginning by introducing readers to the rural energy situation in each country before explaining the history of each project, its benefits, its challenges, and a brief conclusion.

The next chapter (Chapter 2) introduces readers to the technical dimensions of solar home systems, biogas digesters and gasifiers, household wind turbines, microhydro dams, and improved cookstoves. It then elaborates on the contributions these technologies can make to household energy security and community wellbeing.

Chapter 3 recounts the phenomenal success of the non-profit company, Grameen Shakti, which has installed almost half-a-million solar home systems, 132,000 improved cook stoves, and 13,300 biogas plants among 3.1 million beneficiaries in a remarkably short period of time. They plan to ramp up their expansion still so that by 2015, more than 1.5 million SHSs are in place throughout the rural floodplains of Bangladesh, along with 100,000 biogas units and five million cookstoves. This chapter briefly explores the history of the Grameen Shakti and then identifies six distinct benefits that have made their program stand out: expansion of modern energy access; less deforestation and fewer greenhouse gas emissions; price savings; direct employment and income generation; and improved public health and better technology. The chapter also points out challenges related to staff retention and organizational growth; living standards; technical obstacles; affordability; tension with other energy programs; political constraints; and awareness and cultural values.

Chapter 4 depicts how from 2002 to 2007, more than 400,000 solar home systems were sold in northwestern China under the $316 million World Bank/Global Environment Facility-supported Renewable Energy Development Project (REDP). REDP has been hailed as a best practice example in solar home system deployment, winning the prestigious Ashden Award in 2008 for its unprecedented scale and its combination of technology improvement and market development. It is the World Bank's largest SHS project to date. The chapter shows how solar home systems continue to provide monetary and non-monetary benefits to users and that their portability especially complements the lifestyle of nomadic herders that roam the vast expanses of the country's Northwestern region. However, we also find that even for such highly successful programs, purchasing decisions are still based on price rather than quality, and after-sales service networks remain weak.

Chapter 5 delves into the currently ongoing Rural Electrification Project (REP) sponsored by the World Bank in Lao PDR. This is the fourth such project by the World Bank in Laos, and its total cost is estimated to be $72 million, implemented over eight years (2003 to 2013). The effort is driven by the government's ambition to raise the country's electrification rate to 70 percent by 2010 and 90 percent by 2020. In the project, both grid expansion through hydroelectricity and off-grid technologies such as solar home systems are being deployed, the latter in areas that are too remote and sparsely populated to justify grid access. During the first phase of the REP, which ran from 2006 to 2010, 65,000 households were connected to the grid and another 17,000 were provided access to solar home systems.

Chapter 6 evaluates the Renewable Energy and Rural Electricity Access Project (REAP) in Mongolia, an internationally sponsored $23 million program that distributed almost 50,000 solar home systems and small-scale wind turbine systems to rural energy users, mostly nomadic herders. It first briefly describes the history and current status of the electricity sector in Mongolia as well as the current state of rural electrification and energy use among nomads and off-grid herders. Next, the chapter explains the genesis of the REAP project, describing its three primary components related to herder electrification, expansion of *soum* electricity services and national capacity building. Two sections explore the benefits stemming from REAP as well as its lingering challenges. Benefits included expanded access to energy supply among herders and rural users living primarily in *gers*, collapsible tents; enhanced quality of technology; improved affordability of energy services; and reduced greenhouse gas emissions. Challenges relate to the upfront cost of renewable energy systems; dependence on imported technology; remaining gaps in institutional capacity; poor public awareness; and a political commitment to fossil-fueled and centralized energy systems.

Chapter 7 explores the implications of the Rural Energy Development Project (REDP) in Nepal, a scheme funded by a consortium of multilateral donors, including the World Bank and the Government of Nepal, to promote off-grid microhydro energy. The most successful part of the project focused on Microhydro Village Electrification and distributed more than 250 units to 50,000 households in less than ten years. The chapter highlights how community involvement and mobilization; institutional diversity; reliance on simple technologies matched in proper scale to energy end-uses; maintenance and after sales service; flexibility; and restructuring can provide a winning formula to enhance the effectiveness of energy development programs, even in times of conflict or civil unrest.

Chapter 8 documents the successes of Sri Lanka's Energy Services Delivery (ESD) program which ran from 1997 to 2002 and was funded by the World Bank and the Global Environment Facility. The chapter begins by describing Sri Lanka's past and present energy challenges, two years after the end of a protracted 26-year civil war. It then goes on to explain how the program was able to successfully meet all its targets for distributing SHS units and small-scale hydroelectric dams, some of them connected to the national grid, others serving village-scale micro-grids. Subsequent sections explicate the benefits of the program and remaining challenges.

Chapter 9 introduces readers to our first failure. The Village Energy Security Program (VESP) in India ran from 2004 to 2009, and it supported biomass gasifiers, biogas systems, and other hybrid technologies to generate electricity for domestic and productive use. It intended to develop 79 community-scale projects constituting about 700 kW of installed capacity. It also aimed to provide energy for cooking through community and/or domestic biogas plants fueled by animal or leafy waste. However, many of the test projects took a long time to implement, and once implemented, less than half of them remained functional. The VESP has since been discontinued and no new projects have been approved since 2009.

Chapter 10 investigates the Indonesia Solar Home Systems (SHS) Project, funded and administered by the World Bank and the Government of Indonesia from 1997 to 2002. The project aimed to commercialize SHS in rural areas, facilitating the acceptance of the technology as part of a cost-effective strategy of rural electrification. About 200,000 systems were to be installed in areas too remote to be connected to existing power grids. Out of the five provinces originally targeted, the project managed to deploy SHSs in Lampung, West Java, and South Sulawesi and ultimately only 8,054 units were ever sold.

Chapter 11 investigates the Small Renewable Energy Power (SREP) program in Malaysia, the premier policy mechanism implemented by the national government to promote small-scale renewable energy from 2001 to 2010. Eligible technologies included biomass, biogas, municipal solid waste, solar photovoltaics, and mini-hydroelectric facilities less than 10 MW in capacity. The SREP sought to install 500 MW of total renewable energy capacity by 2005, representing about five percent of national electricity capacity. The SREP, however, installed only 12 MW by December 2005, less than 3 percent of its original goal. Malaysian planners therefore modified and extended the SREP for another five years and lowered its target to 350 MW by 2010. However, by the end of December 2010 only 11 projects constituting 61.7 MW of capacity had been installed.

Chapter 12 documents the struggles of the World Bank-assisted Teachers Solar Lighting Project (TSLP) in Papua New Guinea, scheduled for implementation from 2005 to 2010, which aimed to sell 2,500 solar home lighting kits to out-posted schoolteachers while also supporting the growth of local, renewable energy industries. The project was terminated before its target end-date and only one solar kit was sold.

Chapter 13 draws from the case studies to distill twelve key lessons explaining how or why such programs worked, or failed to work. The case studies conclusively show that well-designed and implemented investments in energy access and renewable energy are net beneficial; that is, they produce benefits that far exceed costs. Moreover, the most effective partnerships are those that select appropriate technology, often with input from households themselves; they promote community participation and ownership; and they have robust marketing, demonstration, and promotion activities to inform possible customers as well as policymakers and investors. Furthermore, successful partnerships emphasize after sales service and maintenance, and they actively couple energy services with income generation rather than presuming that households will know how to put such services to productive use. Finally, successful partnerships effectively distribute responsibilities to a variety of institutional actors, strongly emphasize making energy services affordable, build local capacity to create self-sustaining markets, and dynamically adjust targets when things change or go wrong.

Chapter 14 concludes the book and emphasizes that effective programs can rapidly expand energy access and meet millennium development goals simultaneously. Such projects, when designed and implemented properly, can meet national and programmatic targets for electrification and access, sometimes ahead of schedule and below cost. The involvement of women's groups, multilateral donors, rural cooperatives, local government, manufacturers, nongovernmental organizations, other members of civil society, and consumers can increase both the performance and legitimacy of partnerships. Successful renewable energy programs also share not only the rewards of building sustainable markets, but also the risks. They lastly confirm that investments in renewable energy access can benefit households, small enterprises, companies, regulators, and society as a whole.

Chapter 2
The Benefits of Renewable Energy Access

Introduction

Before exploring our ten case studies, it is necessary to first provide some basic details about how off-grid, small-scale renewable energy systems work in the developing world. This chapter shows how a suite of five simple technologies, shown in Table 2.1, namely, solar home systems, biogas digesters and gasifiers, household wind turbines, microhydro dams, and improved cookstoves, can cost-effectively and rapidly respond to pressing energy poverty challenges. This chapter analyzes each in detail before elaborating on the positive social, economic, and environmental benefits they can provide households, communities, and countries.

Table 2.1   **Overview of renewable energy technologies for fighting energy poverty**

Technology	Size	Fuel source	Primary energy services	Cost (levelized US cents/kWh)
Solar Home Systems	10 to 150 Wp	Sunlight	Lighting, communication (radio and mobile phones), television, and small electric appliances	40–60
Biogas Digesters and Gasifiers	6 to 8 m3	Biogas from waste	Cooking and heating	8–12
Household Wind Turbines	0.1 to 3 kW	Wind	Lighting, communication (radio and mobile phones), television, refrigeration, and larger electric appliances	15–35
Microhydro Dams	5 kW to 10 MW	Water	Lighting, communication (radio and mobile phones), television, larger electric appliances, and mechanical processing	5–30
Improved Cookstoves	-	Biomass	Cooking and heating	-

Solar Home Systems (SHS)

The typical Solar Home System (SHS) consists of a solar photovoltaic module, battery, charge controller, and lamp shown in Figure 2.1. Customers in off-grid and rural areas can often choose from a variety of systems and technologies from a 10 Watt-peak (Wp) unit for the poorest households, to a 150 Wp unit for wealthier clients. Larger systems have the capacity to connect televisions, radios, and other electric appliances. Our case studies on Bangladesh, China, Laos, Mongolia, Sri Lanka, Indonesia, and Papua New Guinea involve SHSs.

SHSs offer a very cost-effective way for rural communities around the world to acquire energy services without relying on expensive fossil fuels (such as kerosene) or capital intensive efforts to extend national electricity grids. Costs of yearly battery charging often range from $6 to $15 depending on the type of battery, price of energy, and location. Moreover, overall system costs vary greatly between countries. One assessment of SHSs in China, Kenya, Indonesia, Philippines, Sri Lanka, Brazil, the Dominican Republic, and Mexico found that complete systems cost as little as $10 per installed Wp in China and as much as $100 per Wp in Brazil, differences greater than a factor of ten.[1]

Globally, the World Bank has calculated that 50 Wp and 300 Wp SHSs are already cost competitive in many areas with diesel and gasoline distributed

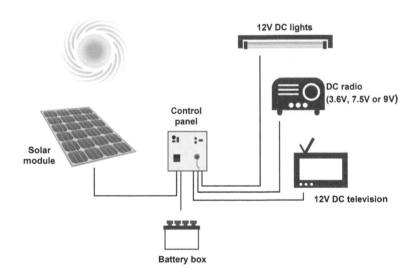

Figure 2.1   Schematic of a solar home system (SHS)

---

1   Miller, D. and Hope, C. 2000. Learning to Lend for Off-Grid Solar Power: Policy Lessons from World Bank Loans to India, Indonesia, and Sri Lanka. *Energy Policy* 28, 87–105.

generators and will see their costs decline, and their advantages to fossil fueled systems improve, even further by 2015.[2] SHS also have immense safety and health benefits—they can displace the use of combusting kerosene, coal, and fuelwood indoors which can lead to higher rates of morbidity and mortality among women and children—and can be easily installed and maintained with minimal amounts of training. Thus, done properly, SHS programs can enhance quality of life, provide a new source of skilled employment for rural technicians, and enhance energy security and reliability due to their decentralized nature.[3]

However, SHS do have some drawbacks. They do not work well at higher latitudes with less sunlight, in tropical areas prone to frequent storms, or on top of foggy mountaintops. Due to the capital expense of purchasing one, they tend to be utilized only by middle- and upper-class rural homes, and they may lead to use for luxuries (powering devices such as televisions) rather than to meet subsistence needs. In some cases they serve only as a "gateway" to electricity coming from the national grid, meaning they do not truly substitute for fossil fuels and conventional electricity systems. Their high value has made them prone to theft and sabotage in some areas—one respondent mentioned they are stolen at the same rate as "automobiles are in rich countries"—and owners and operators must be assiduous in their charging practices for batteries and system maintenance.

Nonetheless, more than 40 national SHS programs exist around the world with more than 1.3 million systems installed at a collective cost greater than $700 million.[4] SHSs thus represent a vital technology employed by multilateral financial institutions in their efforts to curb energy poverty through off-grid electrification, though they do have limitations in that they generally only provide lighting and small amounts of electricity rather than energy for heating or cooking.[5]

Biogas Digesters and Biomass Gasifiers

Biogas is a clean fuel produced through anaerobic digestion of animal, agricultural, and domestic wastes. As Figure 2.2 illustrates, these three forms of organic waste and water typically enter a vessel where they are left to ferment

---

2   World Bank 2007. Technical and Economic Assessment of Off-Grid, Mini-Grid and Grid Electrification Technologies. *ESMAP Technical Paper* 121/07.

3   Miller, D. and Hope, C. 2000. Learning to Lend for Off-Grid Solar Power: Policy Lessons from World Bank Loans to India, Indonesia, and Sri Lanka. *Energy Policy* 28, 87–105.

4   Magradze, N., Miller, A., and Simpson, H. 2007. *Selling Solar: Lessons from More Than a Decade of Experience*. Washington, DC: Global Environment Facility/International Finance Corporation.

5   Karekezi, S. and Kithyoma, W. 2002. Renewable Energy Strategies for Rural Africa: Is a PV-led Renewable Energy Strategy the Right Approach for Providing Modern Energy to the Rural Poor of Sub-Saharan Africa? *Energy Policy* 30, 1071–1086.

and decompose, producing both biogas as well as digested slurry that can be turned into an organic fertilizer.[6] Smaller-scale, two- to three- cubic meter biogas plants tend to be used in homes and communities, suitable for providing gas and heat for cooking three meals a day for an average sized family. Commercial scale systems exist as well, with these larger units offering enough gas to meet the energy needs of neighborhoods, restaurants, tea stalls, and bakeries. These larger systems, installed near large farms, poultry suppliers, and livestock ranches, can supply enough gas for up to 1,000 families. Our Bangladesh and India case studies involve biogas digesters.

By relying on biogas, these units minimize reliance on traditional forms of biomass, animal dung, and charcoal (with their negative environmental and social impacts), and also protect communities from disease by enhancing sanitation. The plants harness gas obtained from livestock and, yes, even human excrement. Biogas systems quite literally have people using their own waste to meet their energy needs.

That said, biogas digesters face some challenges. Once built, they become fixed to a house or structure, meaning if a family decides to move they must leave their system behind. The permanent structure of this type of technology also means if a family or community defaults on their payment plan, banks cannot easily disassemble and repossess the units. They need a sufficient supply of waste to operate optimally, and that amount is far more than a single family can provide (which is why most units tend to be purchased by wealthier families that own

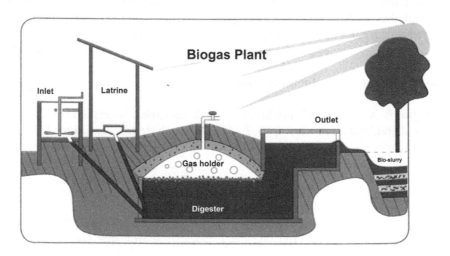

Figure 2.2   Schematic of a biogas digester

6   Gautama, R., Baralb, S., and Heart, S. 2009. Biogas as a Sustainable Energy Source in Nepal: Present Status and Future Challenges. *Renewable and Sustainable Energy Reviews* 13, 248–252.

larger numbers of poultry and livestock). When built improperly or damaged, biogas units can leak methane, quickly erasing any gains made from displacing more carbon-intensive fuels, and the untreated slurry waste has been shown to pollute water supplies. Anaerobic digestion occurs much slower at higher altitudes, meaning biogas digesters work poorly at high elevations, and some women have refused to use them out of the belief they are "unclean" since they are, in essence, powered by human and animal excrement.

Biomass gasification, used in larger systems described in our Malaysia and India case studies, is the process of converting solid biomass fuel into combustible gaseous fuel (called producer gas) through a sequence of complex thermo-chemical reactions. Sometimes this fuel comes from palm oil mill effluent or rice husks; in other cases it is captured from landfills. Our Malaysian case study also includes a refuse derived fuel facility that gasifies and incinerates municipal solid waste.[7]

Household Wind Turbines

Household scale wind turbines operate similarly to their horizontal-access commercial counterparts but in smaller capacities. These devices, shown in Figure 2.3, convert the flow of air into electricity, and are most competitive in areas with stronger and more constant winds, such as locations near the coast or in regions of high altitude. Household turbines generally possess an upwind rotor directly matched to a variable speed electric generator. The modulation of electricity, rotor speed, and orientation are regulated by passive aerodynamic techniques. In some cases, automobile alternators are used to lower cost.[8] Our Mongolian case study involves small-scale wind turbines.

These turbines, ranging from 0.1 to 3 kW, are the second largest technology among our five in terms of capacity, after microhydro dams. With a rotor diameter from 0.5 meters to 14 meters and a battery array, applications commonly include lighting and television in addition to refrigeration and telecommunications. They frequently operate in conjunction with diesel generators or solar PV systems for backup, though they are also the least-used of our five technologies. This is partly because they are more difficult to site, and partly because supply chains and awareness are limited compared to other alternatives such as SHS and microhydro dams.

---

7   Sovacool, B.K. and Drupady, I.M. 2011. Innovation in the Malaysian Waste-to-Energy Sector: Applications with Global Potential. *Electricity Journal* 24(5), 29–41.

8   Practical Action 2011, *Small Scale Wind Power*. Colombo: Sri Lanka.

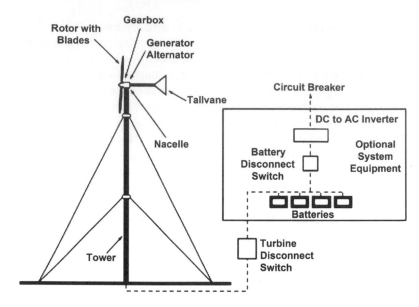

Figure 2.3   Schematic of a small-scale wind turbine

Microhydro Dams

Unlike their larger counterparts which require reservoirs, microhydro dams utilize low-voltage distribution systems and simpler designs that often have a natural river intake, de-sanding basin, masonry lined canal, forebay, penstock, powerhouse, short tailrace, and electronic load controller shown in Figure 2.4. By "micro" we refer to what is commonly discussed as either "mini," "micro," and "small" hydro units from 5 kW to 10 MW.[9] Our Nepal, Sri Lanka, and Malaysia case studies involve these smaller hydro systems.

The most popular systems in Asia use an integrated Pelton turbine and electricity generation unit, which are favored because of their lower cost, ease of installation, light weight (only 35 kilograms for a typical size), and compact design (the entire unit can be transported to a workshop for repairs). Such systems have a developed head of 20 to 100 meters, with diverted flows of about 200 liters

---

9   Small hydropower systems are often divided into five types: "picohydro" units in the range of 1 Watt (W) to less than 5 kilowatts (kW); "microhydro" units from 5 kW to 100 kW; "minihydro" from 101 kW to 999 kW; "small" hydro from 1 MW to 10 MW; "medium hydro" from 11 MW to 50 MW, and "large hydro" above 50 MW. Pico- and microhydro systems tend to serve only individual homes or small villages, and are often for mechanical purposes with electricity as an "add on." Mini hydro systems are typically connected to a national grid or form mini-grids weaving together a handful of villages. Small, medium, and large systems usually serve the national electricity grid.

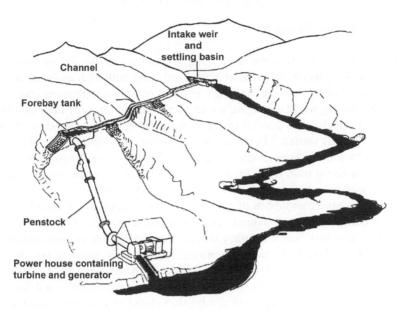

Figure 2.4   Schematic of a microhydro dam

a second. Load controllers are often heavy duty water heaters that act as a sponge to "soak up" extra power to prevent the system from being over loaded. They become essential given that most of the villages do not need power during the day.

Microhydro units have a distinct set of advantages compared to the other technologies in this book. They can provide not only electricity, but also mechanical energy for milling, husking, grinding, carpentry, spinning, and pump irrigation. They are much easier to operate, cleaner, safer, and cheaper than the diesel generator sets they often replace, and local people can be trained to manage them without any technical background in engineering or maintenance. They can also provide electricity in remote mountain areas unsuited for biogas (because fermentation takes more time at higher altitudes) and SHSs (because of consistent fog and cloud cover).

They are, however, not perfect. To work properly, microhydro systems need continuous, dedicated maintenance. Their multi-functionality, the fact that they can perform multiple energy services at once, can become a "curse" when they breakdown or need to be refurbished, leaving communities suddenly without a vital technology for lighting, agricultural production, education, and so on. Larger microhydro units require upstream and downstream communities to cooperate and consent to water rights of way—meaning conflicts between communities can prevent effective deployment—and they can disrupt river flows and degrade fisheries when built improperly. Lastly, the energy produced by microhydro units is not always equitably distributed within communities and villages.

Improved Cookstoves (ICS)

The poorest households in the developing world tend to use simple three-stone fires for cooking wood, agricultural waste, and dung with high moisture content. This results in low efficiencies and high amounts of smoke, with many cookstoves in the developing world averaging ten to 12 percent in terms of their efficiency, meaning as much as 90 percent of the energy content of the wood or charcoal used in them was wasted.[10] In some cases, existing cookstoves can be drastically improved by something as simple as adding a chimney or more insulation around the stove to retain heat, as in Figure 2.5; in other cases, older stoves could be replaced with new stoves with drastic increases in efficiency.

The three basic components of any cookstove are a combustion chamber where wood or charcoal are burnt with air; a heat transfer area, where hot gases actually warm pots and cook; and a chimney which removes hazardous gases outside the cooking area. Though the term "improved" is certainly unobjective, they can take a variety of forms. ICS frequently require a switch away from charcoal or polluted wood to "healthier" fuels such soft biomass, crop residues, and firewood; they have a grate and an improved combustion chamber; and they almost always have a chimney. They utilize higher temperature ceramics, fire resistant material, longer lasting metals, and possess more insulation and a better frame that guides hot gases closer to cooking pots. They can cook more food at once and many have coils around the combustion chamber to heat water while cooking is in process. Some improved stoves are connected to radiators or space heaters so that heat can be recycled and/or vented to other rooms and some stoves send heat through pipes directly into a brick platform that occupants sleep on at night. Other improved stoves are "fuel flexible" and can combust coal and biomass, although doing so requires homeowners to insert a different combustion chamber for each fuel. Improved stoves are also often aesthetically pleasing with beautifully

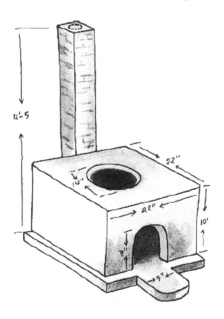

Figure 2.5   Schematic of a one-mouth improved cookstove (ICS)

---

10   Jones, R. 2010. Energy Poverty: How to Make Modern Energy Access Universal? *Special Early Excerpt of the World Energy Outlook 2010 for the UN General Assembly on the Millennium Development Goals*. Paris: International Energy Agency/OECD.

designed tile and artwork, making them something to be proud of and handed down to children, regarded as a family asset.[11]

The most popular models are one-, two-, and three-mouthed clay cookstoves which cut fuel use by half and have chimneys that create a smoke-free cooking environment, improving air quality within the home. These efficient cookstoves not only result in less fuel consumption (typically reducing fuel needs by 40 to 50 percent), they also facilitate shorter cooking times, generate more heat, and reduce indoor air pollution by 20 percent. The technical benefits of these improved stoves, moreover, are manifold. They can be installed quickly, often taking only one to two days. They last longer, with lifetimes of ten years compared to five for traditional stoves. They can be constructed quicker, with prefabricated models taking only seven to 15 days to mold, and they are affordable, costing only $12 for a complete three-mouthed model with a chimney.

As with our other technologies, ICS do have some shortcomings. The most obvious is that the term "improved" is subjective, and it changes over time. An improved stove installed ten years ago is probably no longer an improvement over existing models, and vendors have been known to call stoves "improved" even when they are not. Some of the families we met with expressed a concern that ICS cook food too quickly; that is, they had grown accustomed to the fuel amounts and timing associated with an older stove and became quickly frustrated when the new stove "ruined" their meals. ICS may not meet a family's entire cooking needs—families may wish to boil, bake, and broil with other cooking devices—and they still depend primarily on fuelwood, and many of the burdens associated with its collection and use (though these are substantially less for a truly "improved" stove).

Benefits

As this section shows, despite their flaws these five technologies have the potential to bring massive gains in reducing poverty and heightening productivity, improving health and cutting indoor air pollution, promoting gender equity and education, and preventing deforestation, climate change, and environmental degradation.

Poverty and Productivity

The provision of modern renewable energy services can expand income-generating activities that can greatly reduce poverty, and also help diversify the economies of developing countries against fossil fuel shocks and price spikes.

Poverty and energy deprivation go hand-in-hand, with energy expenses accounting for a significant proportion of household incomes in many developing

---

11   Brown, M.A. and Sovacool, B.K. 2011. *Climate Change and Global Energy Security: Technology and Policy Options*. Cambridge: MIT Press.

countries. Generally, 20 to 30 percent of annual income in poor households is directly expended on energy fuels, and an additional 20 to 40 percent is expended on indirect costs associated with collecting and using that energy, such as health care expenses, injury, or loss of time. In other words, the poor pay on average eight times more for the same unit of energy than other income groups.[12] In extreme cases, some of the poorest households directly spend 80 percent of their income obtaining cooking fuels.[13]

As Table 2.2 shows, the relationship between energy consumption and quality of life is almost monotonic—with the countries with the highest GDPs in the world also having the highest levels of energy access. Lack of energy services, conversely, limits the productive hours of the day for business owners and heads of households, and also inhibits the types of business opportunities available.

Our five technologies, by contrast, make possible a variety of income generating activities, including mechanical power for milling grain, illumination for factories and shops, heat for processing crops, and refrigeration for preserving products.[14] Furthermore, the broader use of renewable energy helps insulate economies from fossil fuel price spikes and diversifies the energy mix, producing significant macroeconomic savings. For instance, countries with underperforming electricity networks tend to lose one to two percent of GDP growth potential due to blackouts, over-investment in backup electricity generators, energy subsidies, and inefficient use of resources.[15]

Dependency on fossil fuels, particularly oil, results in severe macroeconomic shocks. One study looked at the world average price of crude oil for 161 countries from 1996 to 2006, when prices increased by a factor of seven, and concluded that lower-middle income countries were the most vulnerable followed by low-income countries, even though these countries consumed less oil per capita than industrialized or high-income countries.[16] The reason is that the ratio of value of net oil imports to gross domestic product tends to be higher in lower income countries, meaning they spend a greater share of their GDP on energy imports.

---

    12   Hussain, F. 2011. Challenges and Opportunities for Investments in Rural Energy, presentation to the United Nations Economic and Social Commission for Asia and the Pacific (UNESCAP) and International Fund for Agricultural Development (IFAD) Inception Workshop on *Leveraging Pro-Poor Public-Private-Partnerships (5Ps) for Rural Development*, United Nations Convention Center, Bangkok, Thailand, September 26.

    13   Masud et al. 2007.

    14   Eric, D.L. and Kartha, S. 2000. Expanding Roles for Modernized Biomass Energy. *Energy for Sustainable Development* 4(3), 15–25.

    15   United Nations Development Programme 2010.

    16   Bacon, R. and Kojima, M. 2008. *Vulnerability to Oil Price Increases: A Decomposition Analysis of 161 Countries.* Washington, DC: World Bank, Extractive Industries for Development Series.

Table 2.2 **GDP per capita and energy consumption and poverty in selected countries, 2002**

Country	GDP per capita	Electricity consumption per capita (kwh)	Commercial Energy consumption per capita (kgoe)	Population below the national poverty line (%)
United States	36,006	13,241	7,725	-
Japan	31,407	8,203	3,730	-
Korea	10,006	6,632	3,284	-
Brazil	2,593	2,122	717	17.4
China	989	1,139	561	4.6
South Africa	2,299	4,313	2,649	-
India	487	561	318	28.6
Ghana	304	404	120	39.5
Uganda	236	66	26	44.0
Kenya	393	140	96	52.0
Senegal	503	151	128	33.4
Malawi	177	76	27	65.3
Chad	240	12	5	64.0
Ethiopia	90	30	29	44.2
Mali	296	34	18	63.8
Niger	190	41	33	63.0

Source: Masud et al. 2007, 19.

Another study noted that the recently rising oil prices of 2010 and 2011 placed an additional 42 million people in the Asia-Pacific region into poverty.[17] A third study assessed the close connection between rising oil prices and food prices, and documented an almost perfect relationship between the two. Higher oil prices result in rising input costs for agriculture such as oil based fertilizers and fuel for motorized and mechanized equipment, as well as a greater demand for biofuels which then divert agricultural feedstocks to produce fuel rather than food. Both factors create higher food prices, and were responsible for increasing the number of malnourished

17 Kumar, N. 2011. Macroeconomic Overview of the Asia-Pacific Region and Energy Security, presentation to the United Nations Economic and Social Commission for Asia and the Pacific (UNESCAP) Expert Group Meeting on *"Sustainable Energy Development in Asia and the Pacific,"* United Nations Convention Center, Bangkok, Thailand, September 27–29.

from 848 million in 2004 to 923 million in 2007.[18] Our five technologies, by displacing the use of oil, kerosene, and diesel, can ameliorate these negative trends.

Health and IAP

Crisscrossing numerous MDG goals—including maternal health, infant mortality, and disease epidemics—energy poverty has serious and growing public health concerns related to indoor air pollution, physical injury during fuelwood collection, and lack of refrigeration and medical care in areas that lack modern energy services.

By far the most severe of these is indoor air pollution (IAP). Most homes without access to modern forms of energy cook and combust fuels directly inside their home. Burning firewood, dung, and charcoal is physiologically damaging, akin to smoking two packs of cigarettes per day. Almost three-quarters of people living in rural areas and half (45 percent) of the entire global population rely on wood and solid fuels for cooking.[19] Yet, as the World Health Organization (WHO) explains:

> The inefficient burning of solid fuels on an open fire or traditional stove indoors creates a dangerous cocktail of not only hundreds of pollutants, primarily carbon monoxide and small particles, but also nitrogen oxides, benzene, butadiene, formaldehyde, polyaromatic hydrocarbons and many other health-damaging chemicals.[20]

There is a hazardous spatial, temporal, and magnitude dimension to such pollution. Spatially, it is concentrated in small rooms and kitchens rather than outdoors, meaning that many homes have exposure levels to harmful pollutants sixty times the rate acceptable outdoors in city centers in North America and Europe.[21] Temporally, this pollution from stoves is released at precisely the same times when people are present cooking, eating, or sleeping, with women typically spending three to seven hours a day in the kitchen.[22] In terms of magnitude, the quantity of pollution emitted within the home is significantly large. The WHO has set an annual exposure level for particulate matter (PM_{10}) at 50 µg/m,^{3} yet daily peaks during cooking with solid fuels can be as high as 10,000 µg/m.[23]

---

18   Thapa, G. 2011. Food, Fuel, and Financial Crises in Asia and the Pacific Region: Impact on the Rural Poor, presentation to the United Nations Economic and Social Commission for Asia and the Pacific (UNESCAP) Expert Group Meeting on *"Sustainable Energy Development in Asia and the Pacific,"* United Nations Convention Center, Bangkok, Thailand, September 27–29.

19   Legros, G. et al. 2009.

20   World Health Organization 2006, 8.

21   World Health Organization 2006, 8.

22   Masud et al. 2007.

23   World Health Organization 2006

Even when these homes have a chimney and a cleaner burning stove (and most do not), such combustion can result in acute respiratory infections, tuberculosis, chronic respiratory diseases, lung cancer, cardiovascular disease, asthma, low birth weights, diseases of the eye, and adverse pregnancy outcomes; as well as outdoor pollution in dense urban slums that can make air un-breathable and water undrinkable—provoking the WHO to call it "the kitchen killer."[24] Table 2.3 shows the most common, and well-established, health impacts of IAP.

Strikingly, IAP ranks *third* on the global burden of disease risk factors at four percent, coming after only malnutrition (16 percent) and poor water and sanitation (seven percent).[25] This places it ahead of physical inactivity and obesity, drug use, tobacco use, alcohol use, and unsafe sex. The most recent data available, presented in Table 2.4, document that IAP is currently responsible for a shocking 1.6 *million* deaths each year—more than 4,000 deaths per day, or almost three deaths per minute. The cost of this burden to national healthcare systems, not reflected in the

Table 2.3   Health impacts of IAP

Health outcome	Evidence	Population	Relative risk
Acute infections of the lower respiratory tract	Strong	Children aged 0–4 years	2.3
Chronic obstructive pulmonary disease	Strong	Women aged more than 30 years	3.2
	Moderate	Men aged more than 30 years	1.8
Lung cancer	Strong	Women aged more than 30 years	1.9
	Moderate	Men aged more than 30 years	1.5
Asthma	Specified	Children aged 5–14 years	1.6
	Specified	Adults aged more than 15 years	1.2
Cataracts	Specified	Adults aged more than 15 years	1.3
Tuberculosis	Specified	Adults aged more than 15 years	1.5

Note: "Strong" evidence means many studies of solid fuel use in developing countries supported with data from studies of active and passive smoking, urban air pollution, and biochemical and laboratory studies. "Moderate" evidence means at least three studies of solid fuel use supported by evidence from studies on active smoking and on animals. "Specified" means strong evidence for specific age or groups. "Relative risk" indicates how many times more likely the disease is to occur in people exposed to indoor air pollution than in unexposed people. See World Health Organization 2006.

---

24   Jin, Y. 2006. Exposure to Indoor Air Pollution from Household Energy Use in Rural China: The Interactions of Technology, Behavior, and Knowledge in Health Risk Management. *Social Science & Medicine* 62, 3161–3176.

25   Holdren and Smith 2000.

Table 2.4   Global health toll of IAP

Level of development	Deaths in children under the age of 5	Adult deaths	Burden of diseases (thousands of daily adjusted life years)
High-mortality developing (38 percent of the population)	808,000	232,000	30,392
Lower-mortality developing (40 percent of the population)	89,000	468,000	7,595
Demographically and economically developed (22 percent of global population)	13,000	9,000	550
Total	910,000	709,000	38,537

Source: United Nations Economic and Social Council 2010, *New and Emerging Technologies: Renewable Energy for Development*. Geneva: United Nations Commission on Science and Technology for Development.

price of energy, is a whopping $212 billion to $1.1 trillion.[26] Almost all of these deaths occur in developing countries, and more than half are among very young children.[27] Put in perspective, deaths from IAP are already greater than those from malaria and tuberculosis.[28] More worryingly, Figure 2.6 shows that by 2030, deaths from IAP will likely be greater than malaria, tuberculosis, and HIV/AIDS.[29]

The IAP statistics for some particular countries are frightening. In Gambia, girls under the age of five carried by their mothers while cooking have a six times greater risk of lung cancer than if their parents smoked cigarettes and were not exposed to IAP from cooking.[30] From 2005 to 2030, ten million women and children in Sub-Saharan Africa will die from the smoke produced by cooking stoves.[31] One investigation of four provinces in China found that IAP affected every person in a rural home, and also that inefficient combustion of fuel is not the only problem; so is food drying and storage. As it concluded:

---

26   United Nations Environment Programme 2000. *Natural Selection: Evolving Choices for Renewable Energy Technology and Policy*. New York: United Nations. Figures have been updated to $2010.

27   Legros et al. 2009.

28   International Energy Agency 2006. *World Energy Outlook 2006*. Paris: OECD.

29   International Energy Agency, United Nations Development Programme, United Nations Industrial Development Organization 2010, 7.

30   Gaye, A. 2007. Access to Energy and Human Development. *Human Development Report 2007/2008*. United Nations Development Program Human Development Report Office Occasional Paper.

31   Gaye, A. 2007.

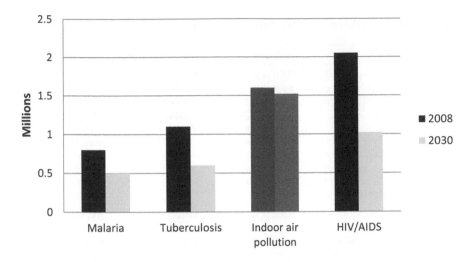

Figure 2.6   Annual deaths worldwide by cause, 2008 and 2030

This comparative study of four Chinese provinces illustrates that exposure to indoor smoke results from interactions of housing, energy technology, and energy use behavior, themselves determined by the diverse socio-cultural and geographical characteristics, and energy infrastructures. Multiple uses of energy—for cooking, heating and purposes such as food drying—result in multiple routes of exposure to IAP.[32]

Table 2.5 overviews some of the study's specific findings.

Unfortunately, IAP is not the only health consequence of energy poverty. Women and children face exposure to health related risks during the burdensome and time-intensive process of collecting fuel. Common injuries include back and foot damage, wounds, cuts, sexual assaults, and exposure to extreme weather. The large number of daily hours women need to collect and use solid fuel leaves them with no other option than to take young children with them, in essence exposing both to the same health impacts.[33]

Countries without access to modern energy also tend to have more dilapidated health systems. Consider that compared to developing countries, infant mortality rates are more than five times higher in energy poor countries, as is the proportion of children below the age of five who are malnourished (eight times higher), the maternal mortality rate (14 times higher), and proportion of births not attended by trained health personnel (37 times higher).[34] In Papua New Guinea, for example,

---

32   Jin et al. 2006.
33   Masud et al. 2007.
34   United Nations Development Programme 1997.

Table 2.5   IAP exposure for four Chinese provinces

	Gansu	Huizhou	Inner Mongolia	Shaanxi
Cooking	Affects primarily women who spend more than 2 hours per day in the kitchen. Some cooking and making tea takes place in living and sleeping areas, affecting older household members.	Affects primarily women who spend 3 hours per day cooking. Cooking the main meal takes place in the common living area, affecting other family members.	Affects all household members since cooking takes place in the same room as living and sleeping.	Affects primarily women who spend 2.5 hours cooking per day. Some cooking and water heating takes place in other living spaces, affecting household members.
Heating	Affects all household members who spend time around the fire pan, especially the children and elderly. Seasonal pattern of exposure is 4–6 months.	Affects all household members who spend time around the heating stove. Seasonal pattern of exposure is 6–7 months.	Affects all household members who spend time on the heated bed. Seasonal pattern of exposure 6–7 months.	Affects all household members who spend time around the ground-stove. Seasonal pattern of exposure 5–6 months.
Food Drying and Storage		Food is stored directly above the stove or in the attic above the chimney outlet, leading to contamination of rice, corn, and chili		

Source: Jin et al. 2006.

many doctors in rural areas without electricity perform births while holding a flaming torch in one hand for light, operating with the other, subjecting pregnant women to severe burns. In Sri Lanka, midwives and those assisting women with giving birth sometimes knock over kerosene lanterns, resulting in "thousands of serious burns" every year.

Furthermore, indirect health effects occur when traditional fuel becomes scarce or prices rise. Meals rich in protein, such as beans or meat, are avoided or undercooked to conserve energy, forcing families to depend on low protein soft foods such as grains and greens, which can be prepared quickly. In other cases, families stop boiling drinking water when faced with an energy shortage.[35]

Luckily, our five technologies can improve general health by enabling access to potable water (wind, solar, hydro), cleaner cooking facilities (biogas, solar electricity, ICS, or hydro), lighting (biogas, solar, hydro, and wind), and

---

35 Murphy, J. 2001. Making the Energy Transition in Rural East Africa: Is Leapfrogging an Alternative? *Technological Forecasting & Social Change* 68, 173–193.

refrigeration (hydro and wind).[36] They can also enable modern preventative, diagnostic, and medical treatment care, including the electrification of healthcare facilities and energy for equipment, sterilization, security, and information and communication technology. Educational awareness raising programs about epidemics and hygiene tend to be enhanced through the modern tools of mass media, such as radios and televisions, which require electricity. The lack of clean water and proper sanitation, a significant cause of disease, furthermore, is linked with lack of access to energy, which can lift subsoil water and sterilize water before use.[37]

Gender and Education

Energy poverty affects both the gender roles within society and the educational opportunities available to children and adults. Gender impacts center primarily on physical injury collecting fuel and the health impacts of IAP (discussed above); expenditure impacts relate to women having to bear the costs of fuel, stoves, and healthcare; and time impacts relate to fuel and water collection, cooking, and the care of sick children. Educational impacts relate to time spent out of school as well as increased absenteeism due to illness.

For instance, women are by large the most vulnerable to energy scarcity; time spent in fuel collection in scarce areas can range from one to five hours per day, frequently with an infant strapped to their back. As the ADB has reported:

> The energy-poverty nexus has a distinct gender bias: of the world's poor, 70 percent are women. Access to and the forms of energy used by a poor community have significantly different impacts on the men and women in it. Existing social and work patterns, particularly in rural communities, place a disproportionate burden of fuel and water collection and their use in the household for cooking on women and girl children, who consequently have to devote long, exhausting hours to this purpose rather than more productive activities, family welfare, or education. However, women's role in decision-making within the household and community is usually very restricted, reducing their say in issues of spending levels and choices, including with respect to energy. This includes the types of fuels used, amounts of energy purchased, the devices and technology chosen, as well as domestic infrastructure characteristics (e.g., stove design, ventilation, etc.). Such decisions are made by the male head of the household, although their burden is borne by the women.[38]

---

36   Larson, E.D. and Kartha, S. 2000. Expanding Roles for Modernized Biomass Energy. *Energy for Sustainable Development* 4(3), 15–25.

37   Masud et al. 2007.

38   Masud et al. 2007.

The labor and time intensity of fuelwood collection, one of these burdens, depends on not only the availability of fuel, but also traveling distance, household size, and season. In the summer months, when wood must be stockpiled for the winter, some women gather firewood twice a day, doubling the amount of time they must allow.[39] In some developing countries, girls spend more than seven times as many hours collecting wood and water than adult males, and 3.5 times as many hours compared to boys the same age. In India, for instance, the typical woman spends 40 hours collecting fuel per month during 15 separate trips, many walking more than 6 kilometers round trip.[40] This amounts to 30 billion hours spent annually (82 million hours per day) collecting fuelwood, with an economic burden (including time invested and illnesses) of $6.7 billion (300 billion rupees) per year.[41]

In addition, current energy production entails occupational hazards that almost uniquely befall women. Women suffer frequent falls, back aches, bone fractures, eye problems, headaches, rheumatism, anemia, and miscarriages from carrying weights often 40 to 50 kilograms, sometimes exceeding their body weight. The energy needs of rural women can be further marginalized if men control community forests, plantations, or woodlots, and if there are other "high value" wood demands on the community that displace their foraging grounds for fuel.[42]

The educational impacts of energy poverty include absenteeism from school as well as increased incidence of illness. Numerous medical studies have documented a strong connection between the effects of IAP mentioned above and acute respiratory infections in children, which is the principal cause of school absences in many countries. School absences arise from such infections, which commonly last seven to nine days each.[43] Moreover, many children, typically girls, are withdrawn from school to complete their chores, including cooking and fuelwood collection. One study noted that literacy levels were lower in fuelwood stressed regions, and it also found a strong correlation between the time children spend collecting fuel and reduced school attendance.[44]

Conversely, our five technologies can help improve both education and gender equality. Table 2.6 depicts a variety of ways they can enhance the status of women by saving time and improving health. In terms of education, one study found that the expansion of energy access improved girl to boy ratios in school, doubling the

---

39   Gaye 2007.
40   Sangeeta, K. 2008. Energy Access and Its Implication for Women: A case study of Himachal Pradesh, India, presentation to the *31st IAEE International Conference Pre-Conference Workshop on Clean Cooking Fuels*, Istanbul, 16–17 June.
41   Reddy, B.S., Balachandra, P., and Nathan, H.S.K. 2009. Universalization of Access to Modern Energy Services in Indian Households—Economic and Policy Analysis. *Energy Policy* 37, 4645–4657.
42   Murphy 2001.
43   Gaye 2007.
44   Gaye 2007.

Table 2.6   Benefits of modern energy services for women

Energy source	Benefits		
	Practical	*Productive*	*Strategic*
Electricity	Pumping water, reduced need to haul and carry mills for grinding, improved conditions at home through lighting.	Increased possibility of activities during evening hours, refrigeration for food production and sale, power for specialized enterprises and small businesses.	Safer streets, participation in evening classes, access to radio, television, and the Internet.
Biomass (improved cookstoves)	Improved health, less time and effort gathering fuelwood, more time for childcare.	More time for productive activities, lower cost of space and process heating.	Improved management of natural forests.
Mechanical	Milling and grinding, transport and portering of water and crops.	Increased variety of enterprises.	Access to commercial, social, and political opportunities.

Source: Masud et al. 2007.

ratio in some districts.[45] Another study of the Philippines noted that the odds of being illiterate are far greater for individuals that lack electrical lighting.[46] Energy services can also enable schools to recruit and retain better-qualified teachers.[47] Lighting from solar and microhydro technologies can extend the time children have to study at night, and can also lead to better equipped schools with computers and the Internet. A similar study documented that almost three-quarters (72 percent) of children living in a household with electricity attended school, compared to only 50 percent of those living without electricity.[48]

Deforestation, Climate Change, and Environmental Degradation

The environmental impacts of energy poverty encompass deforestation and changes in land use, as well as greenhouse gas emissions and black carbon.

---

45   Sovacool, BK, S Clarke, K Johnson, M Crafton, J Eidsness, and D Zoppo. "The Energy-Enterprise-Gender Nexus: Lessons from the Multifunctional Platform (MFP) in Mali," *Renewable Energy* 50(2) (February, 2013), pp. 115–125.

46   Porcaro, P. and Takada, M. 2005. *Achieving the Millennium Development Goals: The Role of Energy Services*. New York: United Nations Development Programme.

47   Porcaro, P. and Takada, M. 2005.

48   Masud et al. 2007.

Since billions of individuals rely on biomass for cooking and heating, about two million tons of it is combusted every day.[49] Where wood is scarce, or populations are dense, the growth of new trees is not enough to match demand for fuel, resulting in deforestation, desertification, and land degradation. Even when entire trees are not felled, the collection of dung, branches, shrubs, roots, twigs, leaves, and bark can deplete forest ecosystems and soils of much needed nutrients.[50] Fuelwood collection can also damage agricultural production: when wood supplies are scarce, people often switch to burning crops—which threatens food security. Moreover, the deforestation and erosion caused by harvesting reduce the fertility of surrounding fields. One recent assessment attributed 6 percent of global deforestation to fuelwood collection.[51]

For example, in Bangladesh trees and bamboo meet about 48 percent of all domestic energy requirements followed by agricultural residues that offer 36 percent, and dung that offers 13 percent.[52] Widespread destruction of forests has occurred to satisfy energy needs, with homestead forest cover reduced to eight percent of its original area[53] and half of Bangladesh's natural forests being destroyed in a single generation by people collecting fuelwood.[54] Similarly, about four percent of China's standing forests are used as fuelwood and roughly 13 percent of cultivated land in China is used to grow fuelwood.[55]

The link between fuelwood collection and deforestation does not hold for all countries, however. One study noted that many times villagers gather not from forests but "invisible trees" which exist not in dense patches but spread around fields, next to houses, and along roads. These trees do not show up on most satellite images of forests or in national forest surveys.[56] Another meta-study of the causes

---

49   World Health Organization 2006.

50   Alam, M.S., Islam, K.K., and Huq, A.M.Z. 1999. Simulation of Rural Household Fuel Consumption in Bangladesh. *Energy* 24, 743–752; Islam, K.R. and Weil, R.R. 2000. Land Use Effects on Soil Quality in a Tropical Forest Ecosystem of Bangladesh. *Agriculture, Ecosystems and Environment* 79, 9–16.

51   Velumail, T. 2011. Regional Context for Improving Access to Energy: Energy for Creating and Sustaining Livelihoods, Presentation to the United Nations Economic and Social Commission for Asia and the Pacific (UNESCAP) and International Fund for Agricultural Development (IFAD) Inception Workshop on *"Leveraging Pro-Poor Public-Private-Partnerships (5Ps) for Rural Development,"* United Nations Convention Center, Bangkok, Thailand, September 26.

52   Miah, D., Al Rashid, H., and Shin, M.Y. 2009. Wood Fuel Use in the Traditional Cooking Stoves in the Rural Floodplain Areas of Bangladesh: A Socio-Environmental Perspective. *Biomass and Bioenergy* 33, 70–78.

53   Miah, D., Al Rashid, H., and Shin, M.Y. 2009.

54   Peios, J. 2004. Fighting Deforestation in Bangladesh. *Geographical*, 14.

55   Chan, M. 2000. *Air Pollution from Cookstoves: Energy Alternatives and Policy in Rural China.* Pittsburg, Pennsylvania: Carnegie Mellon University, 10.

56   Chan, M. 2000, 10.

of deforestation in 152 sub-national case studies found that only in Africa did wood collection seriously contribute to tropical deforestation.[57]

Still, the most comprehensive studies suggest that fuelwood collection is clearly an important contributor to land degradation and deforestation. One study of forest stocks around a sample of 34 cities in the developing world noted that even as these urban areas grow, deforestation occurs. As the study concluded:

> The per capita consumption of biomass fuels persist at a relatively high level until the advanced stages of the energy transition, and the aggregate consumption of biomass fuels does not necessarily decline with income growth. With total biomass energy consumption continuing at a high level as cities develop, the demand pressures on surrounding forested land will continue even after cities have reached the later stages of the modern fuels transition.[58]

Also, the ADB has noted that:

> While deforestation may have many causes other than fuelwood collection that are more important contributors, such as logging, commercial charcoal production, conversion of land to agricultural use, etc., studies show that continued fuelwood harvesting can accelerate such depletion while also diverting biomass away from soil conditioning that can aid vegetative re-growth. Although conditions resulting in deforestation and their underlying determinates may be complex and location-specific, there is little doubt that reducing biomass fuel dependence among the poor … can help relieve the pressure on such natural resources and improve their sustainability.[59]

Apart from its environmental damage, fuelwood-driven deforestation results in two significant social and economic impacts: an increased burden on fuelwood collectors and farmers, and increased fuel prices. First, as stockpiles are depleted, women and children need to travel longer distances to collect fuel, requiring more time and energy. Moreover, such collection typically interferes with the viability of farms and other rural livelihoods that rely on trees for their own income.[60] Second, deforestation results in severe price increases of fuelwood. As deforestation in Bangladesh has accelerated, demand for wood has outpaced supply, causing the price of wood to increase from $0.35 in 1980, to $1.27 in 1991, and $1.69 in 2007 per bunch. When put into the context of a typical household budget, about 50 percent of the annual income of rural households in Bangladesh is now spent on fuel.[61]

---

57   Modi et al. 2005, 30.
58   Barnes et al. 2004.
59   Masud et al. 2007.
60   van der Horst and Hovorka 2008.
61   Biswas, W.K., Bryce, P., and Diesendorf, M. 2001 Model for Empowering Rural Poor through Renewable Energy Technologies in Bangladesh. *Environmental Science &*

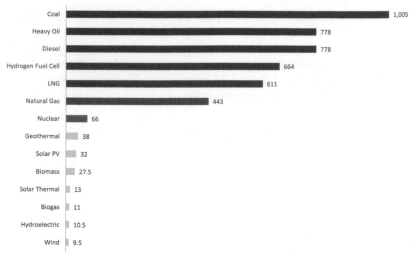

Source: Sovacool, B.K. 2008. Valuing the Greenhouse Gas Emissions from Nuclear Power: A Critical Survey. *Energy Policy* 36 (8), 2940–2953.

Figure 2.7   Lifecycle greenhouse gas emissions for energy systems (grams of CO$_2$e/kWh)

A second environmental impact of energy poverty involves climate change and black carbon. In terms of climate change, burning solid fuels in open fires and traditional stoves has significant global warming effects, due to the release of methane and carbon dioxide.[62] Reliance on biomass fuels and coal for cooking and heating is responsible for about ten to 15 percent of global energy use, making it a substantial source of greenhouse gas emissions.[63] One study, for example, projected that by 2050, the smoke from wood fires will release about the same amount of carbon dioxide as the entire United States.[64] By contrast, when direct and indirect carbon emissions are included, our five technologies are some of the least greenhouse-gas intensive sources of energy, a benefit shown by Figure 2.7. Furthermore, renewable energy technologies not only mitigate emissions, they can also promote adaptation to climate change and a suit of social and economic benefits displayed in Table 2.7.[65]

Lastly, cooking and heating fires are a major source of black carbon (referred to by scientists as "carbonaceous aerosols" and commonly called "soot"), extremely potent contributors to climate change that result from the incomplete

Policy 4, 333–344.

62   Legros et al. 2009.

63   World Health Organization 2006.

64   Gaye 2007.

65   Christensen, J. et al. 2006. *Changing Climates: The Role of Renewable Energy in a Carbon-Constrained World*. Vienna: REN21/UNEP, 28.

Table 2.7   Climate change and development benefits of renewable energy

Type	Application	Mitigation benefits	Adaptation benefits	Social and economic development benefits
ICS	Electricity generation and heat.	Reduced use of charcoal and fuelwood, less pressure on natural resources.	Reduces the likelihood of deforestation and desertification.	Creation of jobs and livelihood opportunities, reduced drudgery, reduction of incidents related to indoor air pollution and respiratory infections.
Wind	Crop processing, irrigation, and water pumping.	Decreased dependence on wood and biomass, avoidance of carbon dioxide emissions.	Greater resilience through reduced vulnerability to water scarcity, more adaptation choices through irrigated agriculture.	Greater prospects for income generation, improved quality of life, reduced risks of vector born diseases, improved water supply and food security, reduced migratory fluxes, improved school attendance (especially for girls).
Biogas	Production of sludge for fertilizer.	Reduced use of pesticides and fertilizers.	Adapting to soil erosion, aridity, and environmental degradation.	Better prospects for agricultural productivity and income generation.
SHS	Cooking, lighting, and water heating.	Reduced consumption of fuelwood, kerosene, and batteries, improved local air quality.	Improved education through illuminated studying and access to information and communication technology.	Improved quality of life as well as better health and sanitation through streetlights and boiled water.
Microhydro	Lighting, agricultural processing.	Reduced greenhouse gases, protection of land cover.	Improved social resilience.	Improved health, greater school attendance.

combustion of coal and wood.[66] Combustion of fossil fuels and wood that occurs at higher temperatures tends to avoid producing black carbon, but low temperature processes such as agricultural burning, wildfires, diesel engines, and cookstoves account for a majority of black carbon emissions. Black carbon particles act like

---

66   Robert E. Koppa and Denise L. Mauzerall, "Assessing the climatic benefits of black carbon mitigation," *Proceedings of the National Academies of Science* (June 21, 2010), pp. 11704–11708.

"tiny heat-absorbing sweaters" that "warm the air and melt the ice" by attracting the sun's heat when they settle in clouds or on surfaces.[67]

While the precise global warming potential of aerosols can vary, in general, black carbon absorbs roughly one million times more solar energy than a unit of carbon dioxide per mass. Researchers from the National Aeronautics and Space Administration in the United States have noted that a single ounce of black carbon absorbs as much sunlight as would fall on an entire tennis court; a pound absorbs 650 times as much energy during its one or two week lifetime as a pound of carbon dioxide would absorb over an entire century.[68] This high amount of "global warming potential" is attributed to the tendency for black carbon aerosols to affect not only cloud coverage but cloud albedo, snow and ice albedo, and optical aerosol mixing.[69] These features make black carbon the second-largest cause of global radiative forcing after carbon dioxide, in essence meaning it is responsible for 15 to 55 percent of *all* global warming.[70] As one study noted:

> When deposited on bright ice and snow surfaces such as glaciers or in polar regions, black carbon particles may cause several more months of warming by reducing the reflection of light.[71]

Some studies have found that soot and black carbon from India have travelled as far as the Maldivian islands and the Tibetan plateau, hundreds of kilometers away,[72] and that black carbon acts as a "regional" pollutant in the lower atmosphere, spreading to disturb tropical rainfall and monsoon patterns across entire continents.[73] Global warming caused specifically by black carbon has already been connected with the accelerated melting of glaciers on the Tibetan plateau, where glaciers have receded 20 percent since the 1960s,[74] and with the premature melting of snow in the Arctic and damage to crops and lower yields in Asia, Africa, and South America.[75]

---

67 Elisabeth Rosenthal, "Third-World Stove Soot is Target in Climate Fight," *New York Times*, April 15, 2009.

68 Adam Voiland, "Black Carbon's Day on the Hill," April 5, 2010, available at http://blogs.nasa.gov/cm/blog/whatonearth/posts/post_1270497028672.html.

69 Koppa and Mauzerall 2010.

70 Voiland 2010.

71 Kandlikar, M., Reynolds, C.C.O., and Grieshop, A.P. 2009. *A Perspective Paper on Black Carbon Mitigation as a Response to Climate Change.* Copenhagen: Copenhagen Consensus Center.

72 Rosenthal 2009.

73 United Nations Environment Program and World Meteorological Organization, *Integrated Assessment of Black Carbon and Tropospheric Ozone Summary for Decision Makers* (Geneva: United Nations, 2011).

74 International Energy Agency, United Nations Development Programme, United Nations Industrial Development Organization 2010.

75 United Nations Environment Program and World Meteorological Organization, 2011.

The good news is that unlike carbon dioxide, and even methane, black carbon stays in the atmosphere for only weeks, meaning conversions to ICS can remove its global warming effects quite quickly.[76] One scientific assessment by the United Nations Environment Program and World Meteorological Organization noted that as part of a synergistic strategy along with methane recovery in the extractive industries, banning of agricultural burning, and the deployment of particulate filters for diesel engines, the dissemination of ICS could avoid 2.4 million premature deaths and the loss of 52 million tons of maize, rice, soybean, and wheat production *each year.*[77]

Conclusion

In short, our five technologies—SHS, biogas, wind, microhydro, and ICS—offer a means of achieving "development first" since they improve the economic status of populations living in rural areas by increasing human productivity and welfare.[78] Conversely, lack of access to electricity and dependence on biomass still produce a series of cumulative negative social and environmental consequences that mire communities and countries in poverty. Even though modern energy can alleviate many of these detriments, its provision is not commonly associated with or integrated into broader development and poverty goals. Most famously, the United Nations has its Millennium Development Goals (MDGs) which consist of the eight goals and eighteen targets.[79] None of these explicitly deals with energy. Moreover, none of the expressly intended investment clusters needed to achieve these goals—such as increasing food output, promoting jobs, ensuring universal access to essential health services, and investing in improved natural resource management—mention electricity or energy specifically.

Nonetheless, modern forms of energy and electricity interrelate seamlessly with each of the eight MDGs, to the degree that they can perhaps be considered a "meta-MDG." As the United Nations Development Program has noted:

> It is clear that energy services have an impact on all of the MDGs and associated targets. Access to energy services facilitates the achievement of these targets. Failure to consider the role of energy in supporting efforts to reach MDGs will undermine the success of the development options pursued, the poverty reduction targets, as well as the cost effectiveness of the resources invested.[80]

---

76   Rosenthal, 2009.

77   United Nations Environment Program and World Meteorological Organization, 2011.

78   United Nations 1954. *Rural Electrification* (Geneva: Prepared by the Secretariat of the Economic Commission For Asia and the Far East, Department of Economic Affairs).

79   Modi et al. 2005.

80   Modi et al 2005, 2.

Because of this strong relationship between energy and the MDGs, some have even gone as far as calling access to modern energy services a "human right."[81] The chapters to come show how this "human right" can be achieved in six successful case studies, contrasted with four case studies that struggled to meet their goals.

---

81   See Shelton, D. 1991. Human Rights, Environmental Rights, and the Right to Environment. *Stanford Journal of International Law* 28, 103–138; Boyle, A.E. and Anderson, M.R. 1996. *Human Rights Approaches to Environmental Protection.* Oxford: Oxford University Press; United Nations Development Programme 1998, *Integrating Human Rights with Sustainable Human Development.* New York: United Nations; and Filmer-Wilson, E. and Anderson, M. 2005. *Integrating Human Rights into Energy and Environment Programming.* New York: UNDP BDP Environment and Energy Group.

Chapter 3
Grameen Shakti in Bangladesh

Introduction

The cockroaches, calls for prayer, cold showers, and repeated power outages at our hotel in Dhaka, Bangladesh, obscured the fact that a few blocks down the road, something exciting and possibly revolutionary was happening in the campaign to fight national energy poverty. At the nearby offices of the Grameen Bank, a nonprofit company, Grameen Shakti (GS), has kept lights glowing and chimneys smoking in rural Bangladesh households since 1996 through affordable renewable energy projects. Taking a leaf from the success of Grameen Bank, GS pioneered the use of microcredit schemes and technical assistance to promote SHSs, small-scale biogas plants, and ICSs. Together, these technologies have contributed toward reducing deforestation, fighting poverty and climate change, providing energy services, and raising awareness regarding the environment and gender roles. Other novel factors in their approach include a focus on matching energy supply with income generating activities, relying on local knowledge and entrepreneurship, utilizing community awareness campaigns, and innovative payment methods including fertilizer, livestock, and cash.

As this chapter shows, GS operated 1,134 offices throughout every district of Bangladesh and in September 2010, had installed almost half a million SHS, 13,300 biogas plants, and 132,000 ICS among 3.1 million beneficiaries, achievements summarized in Table 3.1. They plan to ramp up their expansion so that by 2015 more than 1.5 million SHSs are in place along with 100,000 biogas plants and five million ICSs. These numbers are all the more impressive when one considers that Bangladesh is one of the poorest countries in the world, and also home to a climate perpetually disturbed by natural disasters and subjected to rapid changes in the political environment.

This chapter describes GS's current activities, the contours of its programs, and likely reasons for its success. It also explores the remaining challenges facing GS. We identify six distinct benefits to their programs—expansion of energy access, less deforestation and fewer greenhouse gas emissions, price savings, direct employment and income generation, improved public health, and better technology—before discussing challenges related to staff retention and organizational growth, living standards, technical obstacles, affordability, tension with other energy programs, political constraints, and awareness and cultural values.

The experiences of GS are important because the organization must operate in a country where economic, political, and social factors have hampered energy development. By the most recent estimates, 58 percent of rural households in

Table 3.1   Summary of GS activities and achievements, September 2010

Total Offices	1,134
Number of Districts Covered	64 out of 64 districts
Number of Villages Covered	40,000
Number of Upazilas Covered	508
Grameen Technology Centers (GTCs)	45
Total Beneficiaries	3.1 million
Total Employees	8,400
Solar Home Systems (SHS) Installed	464,520
Improved Cook Stoves (ICS) Installed	132,529
Biogas Plants Installed	13,355
Installed Solar Capacity	23.23 MW
Daily Solar Output	93 MWh
Number of Trained Technicians	6,795
Number of Trained Customers	176,945
Expansion plan for SHS Units, 2011 to 2015	1.5 million
Expansion Plan for Biogas Units, 2011 to 2015	100,000
Expansion Plan for ICS Units, 2011 to 2015	5 million

Bangladesh are officially "energy poor," and lack access to even basic energy services.[1] A staggering 65 percent of all energy needs in the country are met with "non-commercial" forms of wood, twigs, agricultural residues, and other types of biomass.[2] Modern energy fuels such as kerosene, natural gas, and electricity each meet less than one percent of rural household energy needs, and energy consumption is Bangladesh is less than one-tenth the global average, with only 45 percent of the population having access to the national electricity grid (leaving

---

1   Barnes, D.F. et al. 2010. *Energy Access, Efficiency, and Poverty: How Many Households are Energy Poor in Bangladesh?* Washington, DC: World Bank Development Research Group, June, Working Paper 5332.

2   See Kamal, A. and Islam, M.S. Rural Electrification Through Renewable Energy: A Sustainable Model for Replication in South Asia, presentation to the *SAARC Energy Centre*, August 7–9, 2010, Hotel Sheraton, Dhaka, Bangladesh; Islam, S. 2010. IDCOL SHS Program—A Sustainable Model for Rural Lighting, presentation to the *Promoting Rural Entrepreneurship for Enhancing Access To Clean Lighting Conference*, Delhi, October 28; and Asaduzzaman, M., Barnes D.F., and Khandker S.R. 2010. *Restoring the Balance: Bangladesh's Rural Energy Realities.* Washington, DC: World Bank Working Paper No. 181.

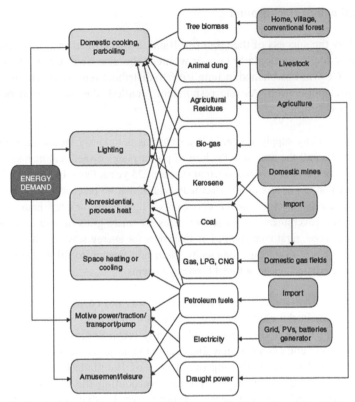

Source: Modified from Asaduzzaman, M., Barnes, D.F., Khandker, S.R. 2010. *Restoring the Balance: Bangladesh's Rural Energy Realities*. Washington, DC: World Bank Working Paper No. 181.

Figure 3.1   Types and sources of energy demand in Bangladesh

96 million people without access, mainly in rural areas) and annual electricity consumption per capita less than 145 kWh per year. National planners admit that they will need at least $10 billion worth of investment and up to four decades before the electricity grid can extend to most rural communities.

Because of this, most Bangladeshis rely on a rich mosaic of energy sources to meet their needs, creating an environment of competing fuels and technologies depicted in Figure 3.1. How GS was able to overcome these barriers to manage some of the largest SHS, biogas, and ICS programs in the world, offers rich insight to those wishing to eradicate energy poverty and provide more equitable access to energy services in other countries. Or, as one of our interview respondents put it, "I really think the GS model can be replicated easily in different countries and for different products."

Energy Access, Poverty, and Development

Description and Background

Inspired by the success of the Grameen Bank, a specialized institution that gives small-scale loans, particularly to poor women who also own nearly 98 percent of the bank, GS was established in June 1996 to distribute renewable energy systems to the rural population. As one respondent recalled, the motivation behind its creation was simple:

> The electricity supply system in Bangladesh is notoriously unreliable, it goes off and on all the time, and the fleet of power plants owned and managed by the government is very old, often in excess of 25 years. Only 40 percent of the population has access to this electricity, and of that 40 percent, most can only afford enough electricity to power one or two light bulbs for a few hours every day. The government has no serious plans for micro-grids or grid extension, leaving poor, rural communities on their own for energy services. We wanted to start a self-sustaining, community-based organization to fill this void and provide efficient, clean, affordable energy systems.

Prior to GS's involvement, renewable energy was not considered viable. As one respondent put it, the challenge was "to develop programs which facilitate rural communities to own and use renewable energy technologies to become eventual partners to bring and expand renewable energy technologies to their communities." GS was created to explicitly improve access of Bangladesh's rural population to renewable energy technologies through overcoming the high upfront cost of installing solar and biogas systems, promoting knowledge and awareness about renewable energy, as well as providing technical training to the rural workforce. Taken from the Sanskrit word for "energy" or "empowerment," GS (which literally means "village energy" or "village empowerment") was kicked off with a $750,000 loan from the IFC as well as a number of small grants.[3]

Though it receives no direct funding from the Grameen Bank, GS is part of a family of Grameen Bank organizations because it adheres to the same principles of empowerment and microfinance, and also since its founding director Dipal Barua was close friends and co-founder of the Grameen Bank together with Nobel Laureate Muhammad Yunus. Its positive ties with the Grameen Bank even led US President Bill Clinton and other world leaders to persuade USAID, the GTZ (now GIZ), and development assistance donors to support the organization with about $6 million in extra grants and loans. Despite this early assistance, however, GS has been operating as a self-sustainable nonprofit company since 2003, and today it receives more than 90 percent of its revenue directly from sales.

---

3   For more on the history of GS, see Barua, D.C. 2001. Strategy for Promotions and Development of Renewable Technologies in Bangladesh: Experience from Grameen Shakti. *Renewable Energy* 22, 205–210; and Gunther, M. 2009. Grameen Shakti Brings Sustainable Development Closer to Reality in Bangladesh, *GreenBiz Magazine*.

As of September 2010, the organization has thousands of employees in Bangladesh, mostly engineers and technicians operating on a variety of employment scales. Its headquarters in the capital city, Dhaka, serves a more political role, informing national policy discussions. However, it has operational management offices in all urban areas in Bangladesh as well as 13 divisional and 130 regional offices, and 950 completely decentralized branch offices. One respondent noted that:

> We don't like to keep things too centralized, meaning branch offices have a great deal of autonomy though every cluster of five or six must report to a single regional office. But we're in the process of decentralizing more autonomy to branch offices because these are the level of our organization closest to the people.

Three of their programs—SHSs biogas, and ICS—have been the most significant.

SHS Program

The SHS program draws on the decentralized nature of solar PV panels to provide electricity to off-grid and inaccessible areas in rural Bangladesh. Systems ranging from a capacity of 10 Wp to 130 Wp are eligible under the scheme, with 50 Wp systems accounting for most sales. The cost of a single 50-Watt panel, typically imported from Japan, along with a battery and associated equipment runs about $380 and can power four lamps and one black and white television, though as Table 3.2 shows, available systems range from a cost of $130 to $940.

Table 3.2   Eligible GS SHSs, 2010

Capacity	Total Load	Operating Hours	Cost (in USD)
10 Wp	Lamp 2 (5W each)	2–3 hours	130
20Wp	Lamp: 2 (5W each) Mobile Charger: 1	4–5 hours	170
50Wp	Lamp: 4 (7W each) Black & White TV: 1 Mobile Charger: 1	4–5 hours	380
85Wp	Lamp: 9 (7W each) Black & White TV: 1 Mobile Charger: 1	4–5 hours	580
130Wp	Lamp: 11 (7W each) Black & White TV: 1 Mobile Charger: 1	4–5 hours	940

The SHS program targets those areas that have little to no access to electricity and limited opportunities to become connected to centralized electricity supply within the next five to ten years. The ease of operating solar panels, their long lifespan, avoidance of combustible fuel, and lack of pollution make them an ideal choice for remote areas, and the program offers microcredit schemes to enable homeowners and businesses to acquire the necessary capital they need to finance installation.[4]

Under the program, interested parties make a down payment to cover 15 to 25 percent of the system cost and then repay GS with a low-interest loan. Given the expense of kerosene and diesel in rural parts of Bangladesh, solar systems typically pay for themselves in three to four years, meaning people that purchase them then own a system that lasts for 17 to 20 further years without fuel costs. Customers can also elect to share in the cost of larger systems under a solar micro-utility scheme that allows shopkeepers or villages to share in making a ten percent down payment spread across 42 months of repayment. GS offers free maintenance while their loans are being repaid, and it trains interested clients in maintenance and operation at no additional cost. As of September 2010, GS had installed more than 464,000 solar home systems, or 23.23 MW of installed capacity, with a production capacity of 93 MWh-peak to more than two million people. About 10,000 clients have availed themselves of the micro-utility financing option, and the entire program was growing at a rate of 20,000 new clients per month.

The program's success rests in part on "active marketing" as well as an emphasis on training and maintenance. In the words of one rural GS employee, "SHSs do not sell themselves, they have to be marketed. We don't get many people coming to our centers on their own and asking for them, so instead we usually go from door to door and ask people directly if they would like to learn more about them." Also, to keep on top of maintenance and the expansion of the SHS program, GS has established 45 Grameen Technology Centers (GTCs) that have trained more than 176,000 heads of households (mostly women) in the proper use and maintenance of solar panels. GS engineers pay monthly visits to households during their financing period, and they offer to extend their services for a small fee of a few dollars per month, afterwards, if a client signs an annual maintenance agreement with GS. The GTCs have educated more than 6,700 women in advanced solar maintenance, enabling them to become full-time specialists who travel around the country to service GS clients for this small fee. Many GTCs actually pay for women to get a four-year technical degree at national universities before they move to rural areas to become residents. As one respondent explained, only women are trained because:

> Bangladesh is still a traditional society, and most families are uncomfortable
> with a man entering any home to interact with women during the daytime. Men

---

4   Komatsu, S., Kaneko, S., Ghosh, P.P. 2011 Are Micro-benefits Negligible? The Implications of the Rapid Expansion of Solar Home Systems (SHS) in Rural Bangladesh for Sustainable Development. *Energy Policy* 39(7), 4022–4031.

work outside of the home in Bengali culture, women work inside the home. If a man goes to a house in the daytime, even to repair or install a SHS, it creates problems.

Apart from providing repairs and maintenance services, GTC-trained women technicians also assemble solar accessories such as lampshades, charge controllers, and inverters, and GTCs run exposure programs for rural school children to increase their awareness about renewable energy.

One interesting and certainly positive byproduct of training women is increased gender sensitivity within communities. As one respondent put it:

> The training of women engineers at GTCs has helped to further gender relations in Bangladesh and partially changed the social status of women. Men seeing women doing things like installing a SHS, building an inverter, refurbishing a lampshade, welding, soldering, or repairing components brings them a new level of recognition in society. What's more, many women become entrepreneurs, start shops, and travel around the region and earn money installing, troubleshooting, maintaining, repairing, and assembling SHS systems and components. We've found that GTCs are a great way to integrate products into rural neighborhoods, and a useful way of showing men that women can earn incomes and be more independent.

The GTC women technicians we met do not always have to work from the office, and are able to procure more than 70 separate components to make charge controllers, mobile phone chargers, DC/AC converters, and inverters on the spot. The women also spoke about earning 300 Tk ($5.81) per customer (or $70 per month if they get a good customer base). One technician told us that "it takes about 5 minutes for me to assemble and inverter or mobile charger and in a day I can assembly dozens of pieces without leaving my home. I can save time, make money, and watch my children all at once."

To further lower costs, GS funnels money back into research and development (R&D) on solar devices and their engineers design and patent original low-cost ballasts and charge regulators that are more efficient and less expensive (often by more than 50 percent) than the ones available on the local market.[5] GS furthermore offers an inclusive warranty that includes a buyback component under which a purchaser can return his or her system to the organization if their area becomes connected to the national grid.

The program has an extremely high satisfaction rate and the users we spoke with seemed genuinely pleased with GS. GS reports that 97 percent of SHS customers payback their loans on time. One woman said that:

---

5   Alamgir, D.A.H. 1999. *Adaptive Research and Dissemination for Development of PV Technology in Bangladesh*. Dhaka, Bangladesh: Grameen Shakti.

> Before we had our SHS, we had to rely on a hurricane lamp and candles for light. My 85 Wp system gives me eight lights but also much more: a mobile phone charger, one fan, and one television. Before, I used 50 liters of kerosene per month, [after], I was able to pay off my SHS in less than 3 years.

Such high payback and participation rates may explain why respondents said that GS "dominates the market," accounting for three-quarters of all SHS sales in 2005 and 2006 and about 60 percent in 2010, dwarfing the second most successful firm with a six percent share, numbers depicted in Figure 3.2.[6] Indeed, one peer-reviewed evaluation of SHS in Bangladesh found that GS was installing as many as 20,000 SHS per month when the closest institutions behind them installed less than 1,000 and the bottom 14 partners had installed a *total* of less than 800 each.[7] As one participant put it, "GS is clearly the most successful SHS program in Bangladesh, possibly even the world."

Biogas Program

The GS biogas program promotes small-scale, two- to three-cubic meter biogas plants to be used in homes and communities. It can also be subscribed to at the commercial scale, with larger systems offering enough gas to meet the energy needs of neighborhoods and commercial enterprises. Researchers have estimated that Bangladesh has more than 30 billion cubic meters of potential annual biogas from livestock and another ten billion cubic meters from human beings.[8]

In 2004, GS started promoting fiberglass biogas units as opposed to traditional brick, sand, clay, and concrete systems. One engineer told us that:

> Fiberglass units cost the same as the traditional ones, but can be constructed quicker and work more efficiently. It takes 15 to 20 days to install a brick biogas system, which is completely impossible during the rainy season. Brick systems also sometimes leak methane from pipes. But fiberglass units can be installed in two to three hours, anytime in the year, and almost never leak.

The financing scheme behind the biogas program is similar to the SHS program. Purchasers pay 25 percent of the total cost of each system as a down payment,

---

6   It became apparent during the review of our chapter that different estimates exist for just how many SHS GS is installing. Some reported numbers show slightly higher annual installations, and the numbers on the GS website show slightly lower numbers than those reported to us by GS managers.

7   Sayadat, N. and Shimada, S. 2007. Analysis of Lifecycle Cost and External Benefits for Grid-Connected Solar Photovoltaic Electricity Generation in Bangladesh. *Bangladesh Journal of Public Administration* 26(2), 109–133.

8   Islam, A.K.M.S, Islam, M., Rahman, T. 2006. Effective Renewable Energy Activities in Bangladesh, *Renewable Energy* 31, 677–688.

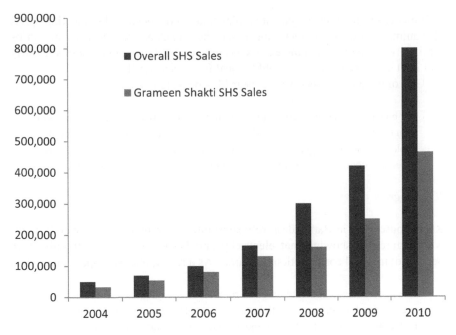

Figure 3.2   **Sales of total SHSs and GS SHSs in Bangladesh, 2004–2010**

and then repay the rest in 24 monthly installments with a 6 percent service charge. Buyers are also encouraged to construct their own plant under the supervision of GS engineers. Biogas plants at the community scale have proven to be quite effective, as many people in rural Bangladesh live as joint families or in joined households with dozens of people close to each other, meaning they can easily share a biogas system. GS manages a special program in Chilmari (the northern part of the country) that provides farmers and communities with the livestock in addition to biogas plants so that they have adequate "fuel." Nationwide, more than 13,000 biogas units were installed by September 2010 and follow up evaluations have found that 90 percent of the plants installed under the project are still in operation and more than 90 percent of the households that use them meet their fuel demand exclusively from these plants.

One interesting offshoot from the biogas program has been the production of high quality organic fertilizer, made as a byproduct from the biogas plant. The use of this fertilizer has reduced the need for chemical fertilizers by 30 to 40 percent in many farms and fisheries, and those not wishing to use the fertilizer have sold it commercially. A three-cubic meter cow dung-based biogas plant can produce eight metric tons of slurry, equivalent to 224 kilograms (494 pounds) of urea and more than 1,200 kilograms (2,646 pounds) of fertilizer, over the course of one year. Mrs Mohammad Abdur Razzak, for example, owns a large chicken farm of 2,000 egg-laying hens and hoses the coop's waste into an underground chamber where it ferments and releases biogas into a pipe that is connected to her cooking stove. Her animals produce so much biogas that she makes an extra $71 per month by renting 10 cookstoves and the excess gas to her neighbors, and the leftover slurry that isn't converted into gas is sold to local farmers as fertilizer.[9]

One of the biogas users we met near Mawna stated that:

> My biogas stove benefits 20 people, it is environmentally friendly, it cooks faster than ordinary wood, and I can cook more meals on it at the same time. I use it to cook rice, fish, and vegetables, typical Bengali fare. It saves me time and money, especially compared to the money I used to spend per month on kerosene.

ICS Program

Rural households in Bangladesh rely substantially on biomass and woody fuels, with Figure 3.3 showing that electricity and kerosene account for only three percent of national consumption by source. As a result, one participant noted that:

> Use of biomass is incredibly high in rural areas. Bangladeshis use biomass for everything, for cooking, for lighting, for constructing houses, for household materials, even for gifts. Since population density is high, the overall impact on forestry in Bangladesh is severe. People think [that] without wood, they cannot survive.

Unfortunately, most rural households depend on traditional cookstoves that are inefficient and often lack chimneys or multiple cooking chambers. One respondent estimated that "these traditional cookstoves use only five to ten percent of usable energy from their fuel, the rest is wasted in the form of heat or smoke."

The ICS program, started in 2006, distributes one-, two-, and three-mouthed clay cookstoves which cut fuel use by half and have chimneys that create a smoke-free cooking environment, improving air quality within the home. Almost all of these stoves are made locally by GS employees, and Figure 3.4 shows the construction of a two-mouth ICS. These efficient cookstoves not only result in less fuel consumption (typically reducing fuel needs by 40 to 50 percent), they also facilitate shorter cooking times, generate more heat, and reduce IAP by 20 percent. GS has so far installed 132,000 ICS and has recently started an ambitious program to install five million of these systems by 2015. It has also trained more than 4,000 local youths and women to manufacture, sell, and repair ICS. A special effort has been made to promote ICS among restaurants, soap manufacturers, and other food providers. Though their claim could not be independently verified, respondents told us that 90 to 95 percent of cookstoves installed under the program are still in use.

---

9   Schroeder, L. 2009. Better lives in Bangladesh – Through Green Power, *Christian Science Monitor*, June 24.

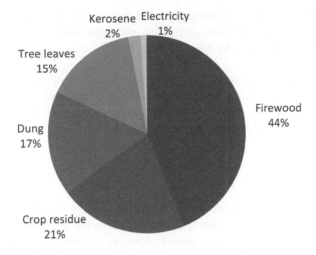

Figure 3.3   Household energy consumption in Bangladesh by source

The technical benefits of these improved stoves, moreover, are manifold. They can be installed quickly, often taking only one to two days. They last longer, with lifetimes of ten years compared to five for traditional stoves. They can be constructed quicker, with prefabricated models taking only 7 to 15 days to mold, and they are affordable, costing only $12 for a complete three-mouthed model with a chimney. Respondents at GS remarked that the ICS is the fastest growing

Figure 3.4   The making of a two-mouthed ICS at a GTC, Singair, Bangladesh

program, with 200 new branches established that year exclusively for ICS as well as the hiring of 500 extra staff. The users we spoke with also said they much preferred the new models. As one stated:

> Before using the ICS I relied on a traditional stove that consumed a lot of wood and made me cough, my new stove has less smoke and dust and cooks food quickly. I can cook for 30 people on my ICS compared to only 15 for my older stove. I can think of no bad things to say about my ICS.

Benefits

GS renewable energy programs have brought at least six distinct benefits to rural Bangladeshis over the past fifteen years. These are the provision of modern energy services; a contribution toward reducing deforestation and greenhouse gas emissions; cost and time savings achieved from more efficient and affordable energy systems; avenues to direct employment and income generation; an improvement in household and public health; and the enhanced availability and accessibility to good quality renewable energy equipment.

By far the most significant benefit from GS programs has been the provision of modern energy services to segments of the population that would not otherwise enjoy them. Only a small fraction of the rural population of Bangladesh has access to the national electricity grid, and the solar panels provided by GS light not only homes and offices but also schools, mosques, and fishing boats; they provide electricity to operate radios, mobile telephones, electric fans, and computers; and they enable equipment such as soldering irons, drill machines, water pumps, and battery chargers. Solar electricity also avoids the use of *kupi* lamps fueled by kerosene, a fire hazard. What is interesting about this expansion of access is that it is demand driven. As one participant noted, "In Bangladesh the electricity grid is so poor even connected homes sometimes want biogas or SHS units. Every household wants some type of renewable energy in place. Because of the energy crisis people themselves try to fill in a need that neither the government nor the market is meeting."

Part of the explanation appears to lie in the fact that the Rural Electrification Board's plans for grid expansion have "stalled" and "failed," according to some respondents. Another went so far as to call government electrification plans "legalized corruption" for the way its tendering and investment decisions are opaque and "enrich construction companies and friends of the political party in power, not people." In contrast to these programs, respondents described GS as "created to exclusively promote renewable energy in rural areas with homegrown technology, unlike other donor organizations, which have business interests in promoting Western technology" and it is considered "the obvious choice and best partner to distribute energy technologies."

Another class of benefits relates to the environment, particularly in the context of deforestation and climate change. Bangladesh is a biomass-centered energy system[10] and widespread destruction of forests has occurred to satisfy energy needs. Homestead forest cover has been reduced to eight percent of its original area and half of Bangladesh's natural forests have been destroyed in a single generation by people collecting fuelwood.[11] Or, as one respondent put it succinctly, "there is no doubt that fuelwood collection causes deforestation, and cooking and energy use accelerates the Bangladeshi deforestation process." Even when fuel collectors avoid cutting down trees and instead take only fodder and kindling, they can still devastate forests by removing key nutrients from the ecosystem.[12]

Cleaner burning ICS directly mitigate deforestation because they double or treble the fuel efficiency of cooking. The ICS program therefore keeps bamboo shoots, rice husks, trunks, branches, shrubs, roots, twigs, leaves, and trees in the fields and forest.[13] The use of electric rice parboilers, electric kettles, and electric stoves, when attached to larger SHS units, can further displace the need for fuelwood collecting. Moreover, the slurry produced from dung and biogas production can help produce local fertilizers that displace the need for those made from fossil fuels.

These benefits not only contribute toward saving Bangladeshi forests, they also cut emissions and earn Bangladesh carbon credits under the Clean Development Mechanism of the Kyoto Protocol. As Table 3.3 shows, GS programs save about 223,400 tons of carbon dioxide per year. An independent World Bank evaluation noted in 2011 that SHS penetration in Bangladesh, due largely to the GS program, has "accelerated the transition to low-carbon economic development" and brought about "significant carbon benefits." [14] Using data from an extensive national household survey, the study calculated that net reductions in national kerosene usage, caused by the diffusion of SHS units, reduced national emissions by 4 percent and electricity sector emissions by 15 percent.

Furthermore, these forestry and carbon benefits exclude other positive social benefits to Bangladeshi society such as fewer hazardous pollutants associated with electricity generation and improved grid efficiency. Two economists calculated that SHS have an additional 9 Tk (10.8¢) per kWh of benefits currently not priced by the existing Bangladeshi market.[15]

---

10   Miah, D., Rashid, H.A., Shin, M.Y. 2009. Wood Fuel Use in the Traditional Cooking Stoves in the Rural Floodplain Areas of Bangladesh: A Socio-Environmental Perspective. *Biomass and Bioenergy* 33, 70–78.

11   Peios, J. 2004. Fighting Deforestation in Bangladesh, *Geographical*, 2004, 14.

12   Islam, K.R., Weil, R.R. 2000. Land Use Effects on Soil Quality in a Tropical Forest Ecosystem of Bangladesh, *Agriculture, Ecosystems and Environment* 79, 9–16.

13   Alam, M.S., Islam, K.K., and Huq, A.M.Z. 1999. Simulation of Rural Household Fuel Consumption in Bangladesh. *Energy* 24, 743–752

14   Wang, L. et al. 2011. *Quantifying Carbon and Distributional Benefits of Solar Home System Programs in Bangladesh.* Washington, DC: World Bank Policy Research Working Paper 5545.

15   Sayadat, N. and Shimada, S. 2007. Analysis of Lifecycle Cost and External Benefits for Grid-Connected Solar Photovoltaic Electricity Generation in Bangladesh. *Bangladesh Journal of Public Administration* 26(2), 109–133.

Table 3.3   Environmental benefits of GS programs, 2009

Program	Yearly fuel/ biomass consumption ton/ liter per unit	Yearly total fuel/biomass consumption ton/liter	Total savings of money (in million $)	Emission reduction per unit (tCO2/Yr)	Total CERs (tCO2/Yr)
SHS	108 l	45,900,000 l	29.3	0.232	98,600
Biogas	1.8 t	22,500 t	1.6	2.08	26,000
ICS	1.8 t	85,000 t	6.1	1.04	98,800
Total			37.0		223,400

GS programs also result in price savings and improve the affordability of energy services. As deforestation in Bangladesh has accelerated, demand for wood has outpaced supply, causing its price to increase from $0.35 in 1980 to $1.69 in 2007. When put into the context of the typical household budget, about 50 percent of the annual income of rural households can be spent on fuel.[16] As the calorific value of these fuels is low, they also require large volumes and therefore force people to spend up to five hours a day collecting fuel that they could otherwise spend making money or gaining a formal education. One woman we spoke with said that she spent two hours every day just collecting cow dung. Furthermore, a study of rural fuel use in the Noakhali district of Bangladesh documented that 40 percent of families collected wood from their own homesteads and 13 percent from the market, but 47 percent depended on them both, meaning they spent both time and money meeting their fuel needs.[17] And fuelwood is not the only expense. Kerosene, used for lighting, can cost up to $0.70 per liter, and most rural families will consume half a liter every night (or as much as $11 every month). Apart from their actual cost, Table 3.4 shows that most families in non-electrified homes can walk as far as 40 kilometers to collect gas, coal, or kerosene, and as far as 30 kilometers to purchase bamboo roots and 35 kilometers to purchase diesel.

Yet another benefit of GS programs is direct employment and income generation. Direct employment comes from GS itself, which now has a full time staff of 8,400 employees, more than half of them Bengali. One respondent noted that "GS has really jumpstarted jobs in the entire sector, by my count more than 12,000 people outside of GS are now employed in areas directly related to GS programs, mostly distribution of equipment and maintenance." Vigorous GS training programs

16   Wahidul, K. Biswas, P.B., and Diesendorf M. 2001. Model for empowering rural poor through renewable energy technologies in Bangladesh, *Environmental Science & Policy* 4, 333–344.
17   Miah, D., Rashid, H.A., Shin, M.Y. 2009. Wood Fuel Use in the Traditional Cooking Stoves in the Rural Floodplain Areas of Bangladesh: A Socio-Environmental Perspective. *Biomass and Bioenergy* 33, 70–78.

Table 3.4 **Distance of nearest source of energy fuel in Bangladesh (km), 2009**

	Electrified Villages				Non-Electrified Villages				Total			
	Mean	SD	Min.	Max.	Mean	SD	Min.	Max.	Mean	SD	Min.	Max.
Bamboo Roots	0.81	2.51	0.00	30.00	1.37	2.79	0.00	10.00	0.94	2.59	0.00	30.00
Diesel	1.24	2.86	0.00	35.00	1.86	2.82	0.00	12.00	1.38	2.86	0.00	35.00
Dung Cake/ Stick	0.68	1.42	0.00	9.00	1.24	2.04	0.00	10.00	0.80	1.59	0.00	10.00
Electricity for Commercial Use	0.59	2.56	0.00	25.00	4.00	9.94	0.00	37.00	0.80	3.54	0.00	37.00
Electricity for Domestic Use	0.55	2.51	0.00	25.00	4.21	10.17	0.00	37.00	0.76	3.50	0.00	37.00
Electricity for Industrial Use	0.61	2.54	0.00	25.00	4.45	9.89	0.00	37.00	0.86	3.63	0.00	37.00
Electricity for Irrigation Use	0.62	2.80	0.00	25.00	3.81	9.73	0.00	37.00	0.83	3.74	0.00	37.00
Firewood	0.77	1.37	0.00	9.00	1.68	2.42	0.00	12.00	0.98	1.71	0.00	12.00
Jute Sticks	0.66	1.29	0.00	8.00	0.97	1.95	0.00	10.00	0.73	1.47	0.00	10.00
Kerosene (White/Blue)	0.32	0.75	0.00	4.00	0.96	1.85	0.00	10.00	0.46	1.13	0.00	10.00
LPG (12 Kg/15 Kg)	5.29	5.60	0.00	30.00	8.35	7.51	0.00	38.00	5.66	5.93	0.00	38.00
Natural Coal	7.00	9.58	0.00	42.00	18.57	11.94	5.00	40.00	8.23	10.39	0.00	42.00
Natural Gas	5.05	7.28	0.00	30.00	9.45	7.69	0.00	20.00	5.41	7.39	0.00	30.00
Octane	4.30	5.11	0.00	36.00	7.34	7.30	0.00	35.00	4.88	5.71	0.00	36.00
Petrol	2.61	3.69	0.00	35.00	4.01	3.87	0.00	20.00	2.91	3.77	0.00	35.00
Rice husk/ bran	0.68	1.45	0.00	12.00	1.49	2.46	0.00	12.00	0.86	1.76	0.00	12.00
Wood/Coal	2.18	3.19	0.00	22.00	4.18	4.50	0.00	20.00	2.55	3.54	0.00	22.00
Total	1.62	3.71	0.00	42.00	2.98	5.36	0.00	40.00	1.85	4.07	0.00	42.00

Source: Chowdhury, B.H. 2010. *Survey of Socio-Economic Monitoring & Impact Evaluation of Rural Electrification and Renewable Energy Program.* Dhaka: Rural Electrification Board.

provide education to entrepreneurs wishing to form businesses related to solar panels, biogas plants, farming and fertilizer, and cooking. One family used their SHS to keep one light for themselves and rent out lights to their neighbors. We saw others utilizing their SHS to charge mobile phones, start commercial enterprises at the household scale, create studying areas for students at night, and pump water for farmers. Indirect benefits accrue to communities as well. Improved lighting enhances educational activities such as reading and writing, extends the working hours in markets to beyond dusk, and powers computers and mobile telephones that enable communication and enhance skills.

One other benefit relates to public health. Reliance on biomass combustion for cooking and lighting produces a significant quantity of hazardous pollutants, including fine particulates, nitrogen oxide, and carbon monoxide, which are typically emitted within the home. One interview respondent compared this type of exposure to "living within a giant cigarette," and noted that homes reliant on biomass tended to have higher rates of acute respiratory infections, eye problems, low birth weights, and lung cancer. More than 50 percent of women in one survey of rural Bangladesh reported headaches, lung disease, asthma, *and* cardiovascular disease related to cooking with biomass fuels.[18] Fuel collection also presents a health hazard, with many women carrying more than their weight in fuel hundreds of kilometers per month. These adverse health effects can lock families into poverty, as they increase expenditures for medical care while also diminishing productivity.[19] Kerosene lighting, too, can endanger health when combusted indoors as it can release carbon monoxide and particulate matter.

Finally, GS has improved the quality and availability of renewable energy equipment. One respondent argued that:

> Before GS began operating in 1996 there was hardly any sort of market for SHS or their components, for biogas or for ICS, but now high quality inverters, charge controllers, and batteries are all made locally, as well as the biogas units and cookstoves. Three companies are setting up manufacturing facilities to assemble SHS here in Bangladesh, and we have 30 suppliers of equipment from 20 countries operating with offices and production facilities near Dhaka, seven manufacturers of batteries alone.

As another respondent noted:

---

18   Miah, D., Rashid, H.A., Shin, M.Y. 2009. Wood Fuel Use in the Traditional Cooking Stoves in the Rural Floodplain Areas of Bangladesh: A Socio-Environmental Perspective. *Biomass and Bioenergy* 33, 70–78.

19   Wahidul, K. Biswas, P.B., and Diesendorf, M. 2001. Model for Empowering Rural Poor through Renewable Energy Technologies in Bangladesh. *Environmental Science & Policy* 4, 333–344.

It used to take two to three months for an ordered SHS to reach Bangladesh, and then another 20 days to send components or workers from Dhaka. Now that GS has centers and offices near practically every community, it takes less than a day to fill an order and as little as two to three hours to send components or technicians.

Challenges

Looking ahead, there are seven disparate challenges that GS must confront should it wish to continue expanding its businesses and fulfill its social objectives. There are challenges related to staff retention and organizational growth; accommodating higher expectations from customers; technology development; reaching lower segments of the population; and competing with other renewable energy programs and grid expansion plans. There is also the issue of political constraints that may result from Bangladesh's often-volatile changes in government as well as lack of communication between different energy stakeholders. Cultural barriers and lack of familiarity with the technologies promoted could also serve to hamper the growth of GS programs in the future if not addressed properly.

The first is staff retention and organizational growth. The structure of GS has grown more complicated as the organization has expanded, and some participants mentioned problems related to job satisfaction. One respondent noted that "like most other organizations in Bangladesh, GS has a high dropout rate. We lose about 30 percent of our staff each year, and we had to hire about 5,000 staff in total for 2010 alone. People don't think that becoming an ICS technician is a dignified job, with no transferrable skills, and many men are reluctant to embrace an organization associated with women's empowerment." Yet another mused that "GS has grown too fast" and that "we're having different types of problems now related to growth, I mean last month we installed a record 34,000 SHS in one month and are expecting to surpass 50,000 per month in 2011, but it's a big question mark as to how long we can sustain that type of distribution." Within certain circles in Dhaka, there are talks about how the quality of service provided by GS has started to deteriorate, as the number of staff cannot keep up with the organization's expansion. A respondent mentioned that GS is particularly gaining notoriety for its poor post-sales services, which became the reason why new customers chose to sign up for SHSs provided by his NGO rather than GS. "By keeping it small, we are able to maintain a quality of service that GS cannot," he said.

A second challenge concerns higher expectations and living standards associated with the employment and income generation GS provides, which often lead to greater material aspirations and rates of consumption. It is one thing to use solar and biogas power to light schools and cook meals, but quite another to power DVD players and support hoards of commercial livestock. GS provides technology and enables people to escape energy poverty, but it does not ensure that they become sustainable stewards of our planet. It empowers them to create their own lifestyle, which can be environmentally sustainable or not. As one respondent argued:

> GS inflates expectations faster than they distribute technology. Once people get a SHS, for example, they want more services, it doesn't stop with lights, they want a television, or if they already have a television, they want a color one, or if they have a lamp, they now want two, and they want them bigger and brighter. In this way, a SHS is like a drug, it gets a household addicted to modern energy services, to convenience and comfort, but cannot always provide the energy needed to back that addiction. Ultimately only the grid may be able to. In this way, GS technologies are not really alternatives to conventional electricity supply, just a temporary bridge until it arrives.

Another respondent joked that households use SHS funded by GS "for soap operas and sitcoms, for entertainment rather than education." To be fair, though, GS rigorously defends their position by noting that livestock can improve food security and televisions can be used to spread knowledge and warn populations before and during natural disasters.

Technology stands as a third challenge. Even though Bangladesh has now started producing components and developing homegrown technologies, their efforts have not been fast enough according to some respondents. While solar panels in Bangladesh are expected to reach grid parity soon, better batteries, warranties on imported solar panels, recycling and disposal of panels, and their operation during longer periods of reduced sun during the monsoon season are all of concern. One respondent commented that even though domestic technology is quite good, many customers in the country refuse to believe it. They are "unwilling to invest so they end up buying a more expensive American, German, or Japanese system because they associate those countries with advanced technology." Another critiqued that "the quality of the batteries used currently in the SHS program is very bad but there is no effort to identify or create better products. As a result, certain battery companies [that] have a monopoly of the market are complacent."

Biogas units, similarly, respondents noted "can leak methane and generate excess gas," "produce a slurry that when not converted into fertilizer can pollute water," and "are permanent [as] they cannot be relocated if a family decides to move." Digesters must also be "continuously fed in order to function properly," "once any fractures occur in the structure it becomes close to useless," and "the process of installation is labor intensive and if payments are not made in a system, it is impossible to repossess because they are fixed structures." Furthermore, since biogas units operate from methane derived from waste, even small leaks can drastically overwhelm the carbon benefits of displacing biomass. One technician commented that the typical digester in Bangladesh used only 80 percent of its methane, the remaining 20 percent leaking through excess pipes or vented after the digester is full. And because they rely on digestion (a biological process) instead of combustion (a thermochemical process) to produce heat and energy, their performance can vary. One respondent jibed that biogas units were "renowned for being temperamental, even more than my spouse!"

These challenges are complicated in the face of Bangladesh's proclivity to natural disasters. Floods, landslides, and tsunamis are among only a few of the recent events that have either directly damaged GS solar panels, stoves, and biogas units, or indirectly destroyed cultivated and arable land thereby reducing the capital villagers have available to make the down payment for GS technologies.[20] When Cyclone Sidr struck Bangladesh in 2007, for example, an estimated 25,000 SHS installed on the coast—about one out of every five—were initially thought to be destroyed.

A fourth challenge is that GS programs require a substantial down payment for their products that is still beyond the means of the poorest members of many communities. One respondent noted that "there are more poor people in Bangladesh than in all of Sub-Saharan Africa. These people cannot even afford a light bulb, how are they supposed to afford a SHS?" Even the Rural Electrification Board has noted that its grid-connected customers are so poor that 80 percent of its 8.2 million customers buy the minimum amount of electricity permitted each month. Numerous respondents argued that "GS technologies are costly, well above what poor Bangladeshis can make in a year," "are still too high for the poorest communities," and "are more a boutique, middle and upper class item than a tool of poverty eradication." Taken together, these factors provoked one respondent to say that:

> The main problem with the GS business model is that it cannot penetrate beyond a certain level of poverty. GS, quite simply, is not for the ultra poor. There are some places in Bangladesh still so remote a person can spend a whole day walking just to collect kerosene to light a lamp at home for a few nights. These people cannot afford the down payment for a 10 Wp panel.

One respondent also noted that "many homes participating in GS programs are richer than they say they are. They will manipulate the data to qualify for subsidies and assistance." As a sign of the growing boutique status of SHS, urban customers in Dhaka who are grid-connected—and therefore ineligible for GS programs—recently got so upset that they have "thrown stones at GS inspectors" and "demanded SHS at a reduced price for their own homes."

Moreover, while solar panels are applicable practically everywhere in Bangladesh, biogas plants are not. Biogas plants are comparatively capital intensive in Bangladesh since they need at least four to five large animals constantly producing waste (human waste alone is not enough). Such units require lots of water and are usually most attractive only for a middle class of Bangladeshi

---

20 Barua D.C. 2001. *Strategy for Promotions and Development of Renewable Technologies in Bangladesh: Experience from Grameen Shakti.* Dhaka, Bangladesh: Grameen Shakti, 5–7.

villagers that are wealthy enough to afford livestock but still poor enough to not afford liquefied petroleum gas or electricity. Respondents repeatedly mentioned that even the smallest biogas units will only work "where someone has at least half a dozen large cows or 200 chickens and their own poultry farm." The biogas units we visited tended to reside only in wealthier homes that had at least a dozen or more cattle.

Interestingly, an extensive study of electrification in Bangladesh undertaken by the REB found that access to renewable electricity did improve living standards but did not significantly eliminate poverty.[21] Researchers surveyed more than 7,000 households as well as thousands of commercial, industrial, and agricultural firms and looked at a mix of electrified and non-electrified households as well as a mix of those with access to renewable electricity and those without. They found that neither electrification nor SHS did much to eliminate poverty. Instead, they noted that electrification seemed to only enhance the attributes of existing families. In other words, factors like land ownership and access to irrigation played a greater role in determining poverty or escaping from poverty than access to SHS or energy services (though the study did admit households with SHS tended to have almost twice as many income earners within the household).

A fifth challenge is the tension between GS, other programs, and plans for grid extension. Most directly, GS programs must compete with other attempts to distribute renewable energy technologies. The REB, for example, has their own SHS program that involves not purchasing a SHS, but "renting" it, requiring households pay a monthly bill for electricity but never own the actual solar panel. One respondent noted that "in my view the REB clearly sees GS as a competitor." Similarly, GTZ is working on a solar lantern program that is aiming to roll out 100,000 units to rural households but "must complete against the demand for SHS promoted by GS."

Moreover, the greater use of solar panels and biogas plants reduces the need to connect (or at least the profitability of connecting) rural Bangladeshis to the national electricity grid. As such, some GS employees mentioned resistance from the REB. Other participants from outside of GS said that the REB sees many of the GS programs as obstructing their plans to extend the centralized grid into rural areas. Thankfully, however, other parts of the Bangladeshi government appear to recognize the importance of GS programs and support them through tax exemptions, special funds, and a renewable portfolio standard aiming for ten percent of national renewable energy supply by 2020.

To be fair to the REB, also, from a macroeconomic sense, the programs being undertaken by GS have yet to make a significant dent in national energy demand. The utilization of renewable energy remains insignificant when compared to the

---

    21   Chowdhury, B.H. 2010. *Survey of Socio-Economic Monitoring & Impact Evaluation of Rural Electrification and Renewable Energy Program.* Dhaka: Rural Electrification Board.

use of fossil fuels combusted by electric utilities.[22] One respondent said "NGO driven renewable energy in Bangladesh is peanuts, it cannot be counted on. One cannot get mega results with micro systems. We need mega-watt scale projects to extend the grid quickly to rural populations." Another remarked that "the efforts from GS are so far just a drop in the bucket in Bangladesh, which has 6,000 MW of annual demand for energy, yet by my count, GS has so far promoted less than 24 MW, a small piece of the pie."

A sixth challenge relates to political constraints: political parties are always changing in Bangladesh, with "not a single party in power for more than a few years." What has remained constant, however, is "corruption" and "poor capacity within government to address energy issues." One respondent argued that:

> There is a complete lack of government commitment and a refusal to empower institutions in place to carry out the task of delivering energy services. Groups like REB and GS are not given autonomy to truly make a difference. They are watched closely and culled when necessary.

Another problem is that GS planners and government officials "operate in separate spheres and do not talk to each other, hence many good efforts are not coordinated and maximized." Yet another is that "GS must compete for attention with other pressing issues like land tenure, natural disasters, diseases, overpopulation, and pesticide pollution, which all matter more to national leaders than energy."

Many of these political obstacles were identified by an independent UNDP study on energy poverty in Bangladesh.[23] The study concluded, among other findings, that resources within the government remain "skewed towards increasing urban energy supplies" rather than rural ones and that most leaders lack "understanding and appreciation of the relationship between energy and poverty." On the other hand, efforts to promote SHS among the rural population mean that the daunting energy challenges facing the urban poor remain unresolved. Nonetheless, the UNDP warned that energy institutions interacted "little among themselves" and have "non-synergistic" policies that often "contradict each other." It argued for an "urgent need" of improved transparency and accountability in how private sector energy contracts are awarded, and also that the focus of most REB programs remains on "electrification" rather than "livelihood options." In short, these political barriers create what one respondent called "a strong current that GS must swim entirely against in order to succeed."

---

22   See Uddin, S.N. and Taplin, R. 2008. Toward Sustainable Energy Development in Bangladesh, *Journal of Environment & Development* 17(3), 292–315; and Uddin, S.N. and Taplin R. 2009. Trends in Renewable Energy Strategy Development and the Role of CDM in Bangladesh. *Energy Policy* 37, 281–289.
23   United Nations Development Programme 2007. *Energy and Poverty in Bangladesh: Challenges and the Way Forward.* Bangkok: UNDP Regional Center.

Seventh and lastly, participants identified awareness and cultural values as challenges. For SHS, these barriers relate to the inability for most rural household members to accept that they can get electricity from what looks to them like "a shiny table." As one respondent put it:

> There is a huge awareness gap concerning SHS in Bangladesh. Some clients actually visited SHS demonstration units more than forty times before they became convinced they would work as claimed. Most people refuse to believe that solar can provide them with light or energy. For them seeing is believing, sometimes seeing forty times.

Another respondent argued that "the number one challenge facing GS and the promotion of SHS is awareness. Even professionals like bankers and school teachers have no idea how a SHS works or what it is, but these people already know very well about kerosene. Suddenly asking them to switch to solar panels is a huge cultural challenge." As one example, consider what one respondent said concerning resistance to electric kettles, stoves, and rice parboilers:

> In our culture, marital ceremonies center on traditional fuels and energy sources. When a new bride first comes to her new husband's house, for example, it is customary for her to cook *boubat* (bridal rice) for the entire new family, supplying a meal for all relatives. The ceremony involves firewood and it has strong symbolic meaning. There's no firewood if you have an electric cooker or appliance. For that reason those technologies are slow to proliferate.

Perhaps because of such sentiments, the Infrastructure Development Company Limited, an infrastructure bank in Bangladesh that also supports GS and manages the national SHS program, had to "do an extensive promotional campaign including workshops, demonstrations, television ads, newspaper ads, radio ads, big billboards in localities, and a mini docudrama about SHS illustrating the difference between a happy family with a SHS and an unhappy one without" to educate consumers about solar energy.

For biogas units, digesters would work best with pig dung but "Bangladesh is a Muslim country and Muslim households refuse to own or eat pigs." Systems also work optimally with the use of human feces injected alongside cow or animal waste but "people have this idea that human waste is unacceptable, that it's mixing in shit with their food, and could counteract the freshness of their meal." One respondent stated that:

> Seventy to 75 percent of solid household waste is biodegradable, creating a huge opportunity to convert it to energy, yet everybody sees human waste as dirty and vile, as something that cannot be used for cooking, which should be pure, irrespective of creed.

Another noted that "villagers do not mind handling cattle dung, but they consider human waste untouchable and unhealthy." Yet another that "Bangladeshi women pride themselves on having a neat and clean kitchen. Even a poor village house wants a separate room for the kitchen [and] will not allow a single animal inside. So they have trouble using gas that comes from cow dung in this somewhat sacred space."

For ICS, one respondent explained that "people have been cooking with traditional stoves for thousands of years, and those stoves need big pieces of firewood. Many do not think a smaller, more efficient stove—with smaller pieces of wood—can actually cook the same." Yet another respondent elaborated that:

> We've found that households will use the improved stove for picture taking, but then go back to the other one when no one is looking. Sometimes, it may be they don't like the shape or the materials, sometimes they may think it does not sit well in the house, or that the food doesn't taste as good, or that it is too small.

In sum, these types of cultural barriers caused one respondent to remark that "no new thing gets accepted immediately, even when its benefits are economically viable, it can be impeded by other hidden factors, reminding us that congruence and compatibility with culture is also needed."

Conclusion

Even in the face of these barriers and challenges, the performance of GS has been remarkable since the organization's inception in 1996. In one of the poorest countries on the planet, in a constantly shifting political environment, and in cities and coastal areas prone to constant natural disasters, GS has catalyzed explosive growth in small-scale renewable energy technologies. The three technologies promoted by GS—SHSs, biogas units, and ICSs—can be found all over Asia and are offered in Bangladesh by more than 20 companies and 25 NGOs, yet GS has a two-thirds market share. As Figure 3.5 shows, their diffusion of SHS grew from less than 6,800 in 2001 to more than 750,000 at the end of 2011, more than a tenfold increase; distribution of biogas units grew from 30 in 2005 to 25,000 by 2011; sales of ICS expanded from 50 in 2004 to 132,000 by 2011. Depending on the year and technology, 95 to 97 percent of households payback their loans to GS on time. As one respondent put it, "GS is growing fast, but is nowhere near market saturation, in a country of 160 million people, we believe we can continue to grow for decades."

With the right financing structure, emphasis on building local capacity, gender sensitivity, and rapid scaling up, an organization like GS can therefore "become a symbol of national pride, an organization whose name is synonymous with clean energy in Bangladesh." Ultimately, the success of GS underscores that capacity-building can go hand-in-hand with expanding renewable energy (eventually

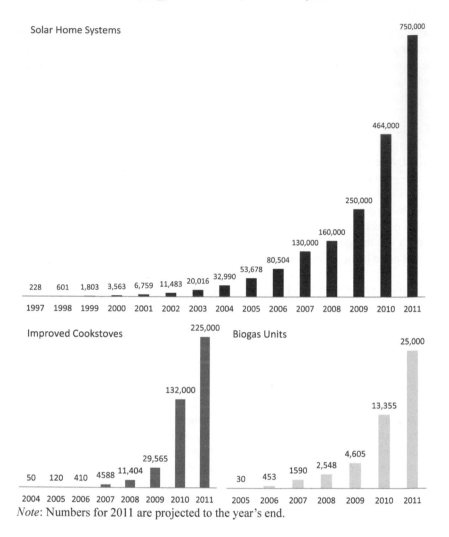

Figure 3.5   Grameen Shakti Sales of SHSs, ICSs, and Biogas Units

improving health and reducing deforestation). GS focuses on training to develop local expertise relating to energy projects, expertise that ensures systems are better maintained but also knowledge that often leads to new entrepreneurs and skills. The effectiveness of GS is based on a simple formula of supplying local energy needs through local yet simple technology constructed and maintained by a community workforce and used to create self-sufficient villages and communities. These systems work best when owned, operated, and repaired by the people themselves.

Chapter 4

The Renewable Energy Development Project in China

Introduction

China has made tremendous strides towards achieving universal electrification over the last 60 years. By 1978, 63 percent of its population had access to electricity, rising to 99 percent in 1998 and now to a present level approaching almost 100 percent.[1] The Chinese clean energy sector is "booming" with more than one million workers and from 2008 to 2010, the price of solar panels within the country dropped in *half* and China now makes 50 percent of the world's solar energy devices.[2] Moreover, China invested $34.6 billion in renewable energy in 2009, more than any other country in the world.[3]

Nevertheless, there are still rural areas where population densities and energy demand are low, rendering grid connections cost-prohibitive and leaving eight million people without electricity.[4] In such areas, renewable energy technologies like wind turbines and SHSs have been identified as least-cost options and therefore the technologies of choice for many rural applications.[5] SHSs in particular are one device that have enabled Chinese households without grid access to use electrical devices for both productive and leisure purposes.

This chapter explores the Renewable Energy Development Project (REDP), a scheme designed to expedite the growth of SHS units in China's off-grid rural areas. At a total sales volume of 400,000 SHSs, the REDP offered opportunities for economies of scale to push down production costs. Being the largest household PV program the World Bank has ever supported, the REDP offers insights into the role of scale in program design.[6] Moreover, the REDP was able to achieve its targets

---

1 Jiahua, P. Wuyuan, L. Meng, W. Xiangyang, W. Lishuang, H. Zerriffi, B. Elias, C. Zhang, D. Victor, *Rural Electrification in China 1950–2004: Historical Processes and Key Driving Forces*, Working Paper #60, Program on Energy and Sustainable Development, Stanford University, 2006.

2 Bradsher, Keith. 2010. "China Builds Lead on Clean Energy." *International Herald Tribune* September 9, p. 1, 16.

3 Plafker, Ted. 2010. "China May Soon Put Its Sunshine to Use." *International Herald Tribune* September, 30, 2010, p. I.

4 IEA, World Energy Outlook Electricity Access Database, in, 2009.

5 A. Zhou, J. Byrne, Renewable Energy for Rural Sustainability: Lessons From China, Bulletin of Science, *Technology & Society* 22 (2002), 123–131.

6 REDP, *REDP Borrower's Report* (World Bank: 2008), p. 46.

in the absence of consumer credit availability and it therefore presents possible lessons for other contexts where credit is tight or microfinance networks non-existent. In addition, the REDP is a well-regarded success story in PV deployment, winning the prestigious Ashden Award for Sustainable Energy in 2008, given by representatives from development organizations including Christian Aid, Oxfam, Climate Care, and Global Village Energy Partnership International.

Description and Background

Several convergent trends convinced Chinese policymakers to design the REDP at the end of the 1990s. A heavy reliance on coal for electricity generation led to serious deteriorations in air quality, especially in urban areas. In 1999, 75 million rural people still lacked access to electricity nationwide. Chinese companies in the solar market were relatively small and inexperienced, with consistent complaints of poor system quality and lack of availability. Little to no consumer credit was available to purchase such systems, and 3 solar companies went out of business in 2001.

Planners designed the REDP to address these prevailing concerns. Of paramount importance at the time was inadequate electricity access among rural households, barriers to private investment in renewable energy manufacturing, and excessive reliance on coal-based power generation.[7] Although formally initiated in 1999, due to implementation delays the REDP "really got going" in 2002. The REDP was broken down into three main components:

- A solar Market Development Support Framework which aimed to distribute 10 MW of SHS (roughly 350,000 units) to nomadic herders and other off-grid households;
- A Technology Improvement scheme intended to improve system quality, decrease costs, and enhance the Chinese manufacturing base;
- A pilot Wind Energy component seeking to build two demonstration wind farms near Shanghai.

Table 4.1 provides an overview of these components. We have excluded the wind energy component from our analysis since it dealt with commercial, utility scale wind turbines rather than household needs, our area of inquiry for this book.

The State Economic and Trade Commission established a Project Management Office (PMO) to coordinate the REDP, which as a result of government restructuring, was transferred to the National Development and Reform Commission (NDRC) in 2003. Consisting of 10 full-time employees assigned to the TI or PV components, as well as Financial and Contracts Management, the PMO was responsible for

---

7   World Bank, *Project Appraisal Document for a Renewable Energy Development Project* (Energy and Mining Development Sector Unit, East Asia and Pacific Regional Office, 1999)

Table 4.1   Overview of Renewable Energy Development Project (REDP)

	PV component	Wind component	TI component
Project management	REDP PMO, under State Economic and Trade Commission (later NDRC)	Shanghai Municipal Electric Power Company	REDP PMO, under State Economic and Trade Commission (later NDRC)
Key objectives	• 10 MW of installed SHS capacity, reaching 350,000 households • Avoided emissions • Reduction in capital costs, measured by $/Wp	• 190 MW of wind capacity, downgraded to 21 MW during 2001 project restructuring • Avoided emissions • Reduction in capital costs, measured by $/kW	• To improve quality and reduce manufacturing costs of PV equipment
Project costs	$96.6 million	$27.08 million	$191.95 million
Key stakeholders	World Bank, NDRC, PMO, PV companies, retailers, end users	World Bank, NDRC, Shanghai Municipal Electric	World Bank, NDRC, PMO, component manufacturers

making all management decisions at the central government level. This included tasks like selecting participating companies, authorizing grant payments, and designating certification procedures for subcomponents. Over the course of the program, they sponsored promotion efforts, like the production of television and film content to expand awareness about renewable energy, and initiated training capacity-building courses and conferences for PV companies.[8] While the PMO was an independent body, their decisions still required approval from the NDRC and the World Bank, with whom they had regular contact.

The initial target areas for the PV component were Inner Mongolia, Gansu, Qinghai, Western Sichuan, Tibet, and Xinjiang, later extended to Shanxi, Ningxia, and Yunnan provinces. As of 1995, more than nine million people were without electricity across these 10 provinces and autonomous regions which range in population density from 0.2–2.3 households per km². Data from Table 4.2 indicates that these western provinces have trailed behind national averages in rural electricity access rates by several percentage points over the described period. Although the REDP did not have an explicit poverty alleviation objective aside from satisfying the energy needs of populations "that would otherwise not receive services,"[9] per capita incomes in the six provinces of initial implementation were 15 to 43 percent lower than the national average in 1997.

---

8 G. Chun, *REDP Promotion and Outreach Report* (GEF/World Bank/NDRC, 2008).

9 World Bank, *Project Information Document: China-Renewable Energy Development Project* (Washington, DC: World Bank, 2001).

Table 4.2   Percent of rural households in China with electricity access

	Northeast	North	Northwest	East	South-Central	Southwest	National Average
1991							80.00
1992							
1993	99.40	96.36	80.96	92.83	88.75	79.49	89.61
1994	99.56	96.67	83.46	95.09	90.82	82.95	91.33
1995	99.68	97.26	85.58	97.10	93.22	84.76	93.30
1996	99.68	97.26	85.58	97.10	93.22	84.76	93.30
1997	99.81	98.04	89.36	98.91	96.07	89.90	95.86
1998							95.86
2000							98.03
2001							98.40

As the sections to follow illustrate, the REDP had a number of distinctive features. Unlike existing projects, such as the Silk Road Program or Brightness Program which focused only on solar panels, the REDP emphasized entire SHSs including lights and components. These other programs had no support for training, capacity building, and the improvement of manufacturing. Under the REDP, SHS designs were modified to make them easier to assemble and disassemble, sized to fit into a metal box made to travel, for nomadic herders. Other programs were geared towards 50 Wp and 100 Wp systems, which were larger and more expensive; the REDP supported units as small as 10 Wp, enough for herders to keep sheep together with one or two lights in a storm, meeting their minimal needs. Lastly, the REDP was "the first" to help companies commercialize and expand their networks, train sales managers, and improve knowledge to organically grow a local market. As one respondent bluntly noted, "other projects are ignorable compared to the REDP." Another argued that "one of the unique elements of REDP was its integrative nature; in this case, the whole of the program was greater than the sum of its parts."

PV Component

Planners configured the REDP to tap into four existing markets for SHSs: (1) first time purchasers, mostly nomadic herders or new homes, who desired smaller capacity systems to be utilized mainly for lighting; (2) greater electricity demand in homes that already had an SHS but wanted larger systems to power more appliances, in essence needing to change from a smaller panel to a larger one; (3) new homes in peri-urban areas that wanted an SHS for backup in the case of blackouts and interruption in grid-supplied electricity; and (4) the global export market, including Asia and Europe.

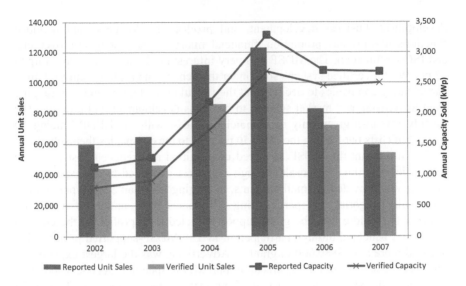

Figure 4.1   REDP's reported and verified SHS sales in China, 2002–2007

The central priorities under the REDP's PV component were to improve product quality, reduce production costs, and install a total of 10 MWp of SHS capacity (originally over the course of 1999–2004, but in actuality from 2002–2007), under the three cost centers of investment, market development, and institutional strengthening. In actuality, the 28 participating companies surpassed the capacity target in 2007 and sold 11.1 MWp of SHS units. The PMO verified a sales volume of at least 400,000, while companies claimed an even higher unofficial figure of some 500,000 SHS sold during the REDP's implementation. SHS units of 10Wp and larger were eligible for subgrant support and the average size of sold units gradually climbed from 18Wp in 2002 to 45Wp in 2007. Annual sales data by unit volume and capacity (kWp) are plotted in Figure 4.1.

Initially, participating companies received a $1.5/Wp subgrant for each SHS sold that passed certification standards. The companies pushed for subgrants to be increased to $2.5/Wp[10], but after the mid-term review in May 2005, they were instead increased to $2/Wp in compensation for the compliance costs incurred by producers in upgrading their equipment to tighter standards. Non-compliance penalties for improper documentation, restrictions implemented after 2005 on grants for high-capacity systems, and the exhaustion of funds led to an average grant size below the nominal rate.

A Market Development Support Facility (MDSF) was introduced in June 2003 and supported activities grouped under the three categories of market

---

10   REDP, Assessment of the Influence of Other Relevant PV Projects on REDP and Suggested Mitigation Actions, in, 2004.

development, business development, and product development, and included functions like product promotion, financial management system improvement, and ISO certification. The MDSF was very interesting: it gave a subsidy of up to $2 per Wp for certified systems sold under the scheme, and it left it entirely up to participating companies to decide what they would spend their funds on, ranging from advertising and market surveys to business development and training. The subsidy was also given only on a matching basis, with the REDP funding 50 percent of costs but leaving the other 50 percent to the companies themselves.

To ensure that the MDSF furthered competition and reduced costs, eligible companies submitted proposals through a process evaluated by an expert team. Support was provided in small amounts, averaging less than $5,000 per activity across all categories and participating companies were under no obligation to publicly disclose how the grant support affected their profits. In the end, 30 companies benefited through participation in 190 MDSF-supported projects that released more than $880,000 in grants.[11] Advertising was the largest expenditure category and accounted for 36 percent of total MDSF spending. As one respondent explained, "the MDSF was so successful the SHS market is now [as of 2010] self-sustaining; it does not need further government intervention to function."

TI Component

REDP planners crafted the TI component to improve the quality of SHSs. The TI component was modeled on initiatives in Western countries and designed to "accelerate technology innovation, with the aim of reducing costs of equipment available in China, while providing high-quality products and performance".[12] As one participant elaborated, "many solar panels sold before the REDP had defects, falsified labels, substandard parts, and quality so poor that they lasted only one to two years." Another remarked that "before the REDP, there was absolutely no quality control and no certification. Households could be buying a lump of rock called a solar panel, with no testing of inverters and controllers, despite there being 20 suppliers on the market."

The TI component therefore included:

- Competitive grants, also on a cost-sharing basis, to investments in technology improvement from component and system manufacturers;
- The creation of a Standards Committee which modified existing standards, developed new ones, and certified testing laboratories;
- The establishment of an "Approved Components" list based on quality tests carried out by the selected testing laboratories as well as a "Testing Team"

---

11   REDP, *The Market Development Support Facility Appraisal Report* (Washington, DC: World Bank, 2008).

12   World Bank, Project Appraisal Document for a Renewable Energy Development Project, in, Energy and Mining Development Sector Unit, East Asia and Pacific Regional Office, 1999.

which travelled throughout China randomly evaluating components;
- The organization of a PV Component Testing seminar for companies whose products did not meet REDP quality standards as well as the publication of quarterly newsletters and bulletins to raise awareness;
- Sponsored visits to trade fairs, testing institutions, and conferences to build recognition about the REDP;
- Sponsored upgrades to PV testing laboratories shown in Table 4.3.

Despite these varying tracks, improving the quality of current SHS products on the market was a core objective of the REDP. To qualify for the REDP support, PV companies were required to demonstrate that all system components complied with prevailing standards. The PMO's Technical Standards Committee outlined all technical specifications for subcomponents, such as the solar module, inverter, controller, battery, and DC lights. Along with a laboratory based at Arizona State University, four centers—Tianjin Institute of Power Sources (TIPS), Post and Telecommunications Industry Products Quality Surveillance and Inspection Center (PTPIC), National Center for Quality Supervision and Testing of Electric Light Sources (NCQSTL), and Shanghai Institute of Space Power Sources (SISP)—were selected to conduct product testing for manufacturing companies with funding available through TI grants.

Other institutes were later added, such as the Photovoltaic and Wind Power Systems Quality Test Center (PWQTC) of the Chinese Academy of Sciences. In addition to testing random product samples in the laboratory, engineers would travel to the western provinces to obtain products for quality inspection and stage spot checks at manufacturers. In this regard, The PMO acted as the intermediary to connect qualified suppliers with PV companies and circulate technical guidelines to national and international companies.

In response to companies applying unapproved REDP labels on their panels, the PMO developed the Golden Sun label, which certified compliance with REDP's standards which were gradually tightened across the project's duration. It was

Table 4.3   REDP upgrades to PV testing laboratories

Before REDP	After REDP
Unable to perform tests on maximum power point tracking characteristics of charge controllers	Equipment installed to test controllers and inverters
Unable to test system efficiency	Able to test any type of controller, inverter, integrated controller and inverter, and battery, including ISO/IEC 17025–2005 accreditation for system tests
Unable to test batteries	
Unable to test hybrid systems	Able to test various configurations of hybrid systems

considered one of the hallmarks of the REDP, which initial project documents did not feature. In 2003, the PMO observed that many subcomponents outperformed the minimum requirements and issued a new national standard. These were again replaced, and in December 2005 the "Solar Home Systems Implementation Standard" took effect and was modeled on the international standard for SHS quality—the IEC 62124—and its local adaptation, GB 9535-1998. The rationale for this standard was to ensure system quality.

Over the duration of the entire REDP, 133 TI supported projects were completed with $2.8 million awarded in funds, 95 percent of them achieving their intended outputs. Some of the specific accomplishments of the TI subcomponent included the improvement and sale of 3.2 million DC lights, 25,000 LED lights, 21,000 SHSs, 75,000 charge controllers, 18,000 integrated controller and inverter systems, 550 hybrid wind-PV systems, 115 PV module laminators, 45 solar simulators, 3 PV array testers, and 1,500 monitoring devices for wind-solar hybrid systems. The TI sub-grants also supported 23 testing and certification projects and contributed to increasing awareness and international collaboration; one individual project sent officials to Germany and the United States to learn about solar PV manufacturing, to Spain to learn about solar thermal water heaters, and to Australia and Tanzania to learn about testing and certification.

Also noteworthy is not what was spent, but what was taken away. Over the REDP implementation period, the PMO extensively fined manufacturers failing to meet quality standards. A total of $1.1 *million* in penalties was levied against manufacturers from 2004 to 2007, numbers reflected in Figure 4.2. To give readers a sense for the differing quality of solar manufacturers, the research team did visit two companies in Xining that were heavily fined as well as two that were not. The ones that were fined featured workers assembling modules practically in the dark, bathrooms with no-flush toilets, leaking ceilings, clouds of cigarette smoke, broken solar panels deposited in corners, and overall poor conditions; the others, by contrast, were well lit, smoke free, air conditioned, and generally clean.

Though we detail its benefits in greater detail below, the end results of the TI component are impressive. Certification of solar equipment is now so good China can sell panels to Europe. As one respondent put it, "without the TI component, I remain convinced that existing private sector research on solar would have proceeded at a snail's pace—now we are at the forefront of international standards."

Benefits

In theory, the main beneficiaries of the REDP were identified as nomads and rural households, people living in areas prone to harmful emissions from coal fired power plants, participating solar companies, and solar technology manufacturers.[13] In practice, the outcomes of the REDP were collectively enjoyed by Chinese PV

---

13   World Bank. 2009.

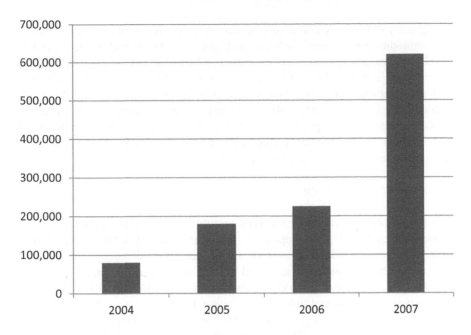

Figure 4.2 **REDP fines and penalties against PV manufacturers in China (U.S. Dollars)**

companies, SHS retailers, and rural end-users, who benefited from technology improvements that reduced costs and improved the quality and standards of the SHS deployed. The REDP was able to reach approximately two million end-users, providing access to modern energy services and improving quality of life through increased income, better family communication levels, increased workable hours, and improved access to information through radio and television. Chinese PV manufacturing companies have also been able to improve the quality of their products and adhere to stricter standards, whereas PV retailers continue to experience an overall increase in sales despite lower annual growth rates. This section divides such benefits into four areas: lower cost and better technology, expansion of energy access, enhanced distribution networks for solar energy, and improved national economic competiveness.

Lower Cost and Improved Technology

The REDP improved solar technology in myriad ways. First, it enabled rural households to move up the "energy ladder" towards larger systems. As Figure 4.3 shows, during the first three years of the REDP, average system size remained about 20 Wp, but it rose to more than double that (45 Wp) in 2007. Second, average unit costs actually declined from $16 per installed Wp to $9 per Wp, though they did slightly rise during the middle of the REDP due to global surges in commodity prices, trends reflected in Figure 4.4. At the end of the project in 2007, panel prices—not SHS prices—dropped below $2 per Wp, one of the most competitive in the world,[14] although our own inspections of retailers in Qinghai saw great variation in prices from store to store, from a low of $1,600 for a 100 Wp unit to a high of $3,000. Still, as one respondent succinctly put it, "After the REDP, prices of solar equipment dropped as the industry improved in technical ways through better products and capacity building, which means manufacturers are already one step ahead of global firms."

Expansion of Energy Access

The population of rural herders is hard to supply with electricity, since, as one respondent exclaimed, "it's impossible to connect moving tents to the national grid." Another remarked that "nomads always need portable energy supply, since they cannot herd in one place and need to keep animals on mountains for food." The REDP quickly filled this niche by expanding access to more than two million people putting more than 400,000 SHS totaling 10 MWp of capacity to use—total system sales amounted to greater than 500,000 and 11.1 MWp, but could not be verified as coming from the REDP. Figure 4.5 illustrates these sales, most of which occurred in the regions of Tibet, Qinghai, Sichuan and Xinjiang. More than 60 percent of these users reported being "satisfied" or "extremely satisfied" to the World Bank's own evaluation team in 2007.[15]

These SHS benefitted households and herders in various dimensions. One was decreased fuel consumption and costs; once they acquired a SHS, their reliance on candles and kerosene declined remarkably. In Gansu and Tibet, kerosene consumption dropped from 19 percent to only 4 percent of all rural households. Another was increased appliance usage, with 91 percent of SHS users purchasing electric lighting systems, 80 percent radios, and 40 percent rechargeable flashlights. The World Bank reports that at the close of the REDP in 2007, 80 percent of systems did not need any repairs during their warranty periods and that 65 percent did not yet need to replace their battery. Most strikingly, 95 percent of REDP beneficiaries reported that their income had increased as a result of their access to electricity.

As part of the project evaluation process, local consultants conducted face-to-face interviews in 2007 with 1,203 households in the Tibetan Autonomous Region and Gansu Province. Of this sample, 69 percent were PV system users and 31 percent were not. Survey questions addressed usage patterns for SHSs and time-use behavior of various members in different households. The evaluation concluded that SHS use had a positive effect on household income for more than 53 percent of respondents. Among other benefits, they also estimate improvements

---

14   World Bank. 2009.
15   World Bank. 2009.

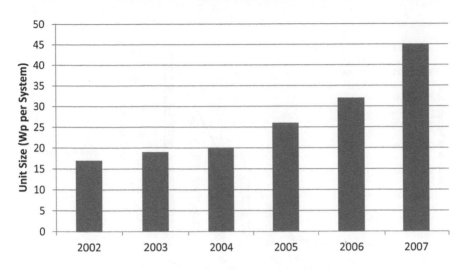

Figure 4.3   Average unit sizes of SHS under the REDP in China, 2002 to 2007

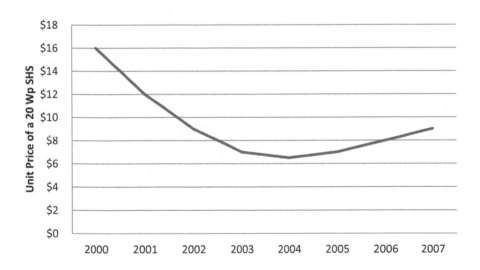

Figure 4.4   Average unit price of a 20 Wp SHS in China, 2000 to 2007

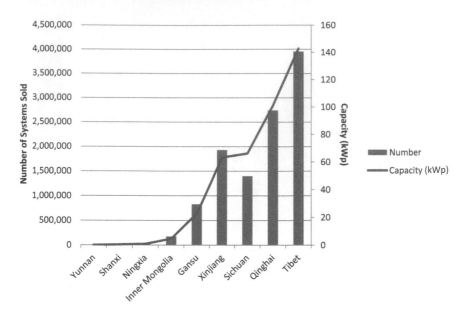

Figure 4.5   Distribution of SHS sales for the REDP in China

in family communication levels, increased workable hours, and improved access to information through radio and television. Use of alternative lighting sources, such as ghee and kerosene lamps, declined as a result of SHS penetration. With the same survey data, the World Bank concludes that "there are strong indications that poverty impacts have been achieved among a considerable number of people."[16]

To supplement the Bank's own evaluation, we interviewed end-users in various parts of Qinghai province, including Dulan, Henan, Nima Te, Zekog, and Zeku Counties. They were primarily nomadic herders with at least one household member spending summer months in the hills. Our interviewees expressed significantly lower interest in listening to the radio and stated greater benefits of SHS ownership through lighting and electricity for mechanical milk separators. Nomadic herders bring SHSs to provide lighting for when they summer in the mountains. There they collect *Cordyceps Sinensis*, an ingredient in traditional Chinese medicine also known as caterpillar fungus and frequently priced higher than RMB 100,000 ($15,800) per kilogram, as well as other herbs. Since they may travel to the mountains for weeks at a time, the SHS provides services matched to their mobility that a fixed grid connection cannot. We also visited brick-constructed winter homes that were grid-connected to local microhydro power projects and in some cases an SHS would provide them with supplemental power. Xining SHS

---

16   World Bank, 2009.

dealers told us that beekeepers also purchase SHS units, usually under 50 Wp, and use them for lighting.

Enhanced Distribution Networks

The REDP augmented and improved existing distribution networks for solar equipment. By 1999, only approximately 90,000 PV systems were in use across China. According to pre-implementation data, the top four companies jointly held a 36 percent market share and most of these companies had been in existence for less than two years. Respondents noted that the REDP, however, "bolstered" and "dramatically strengthened" the Chinese supplier network. We certainly found this to be true during our interviews of five Xining dealers and the sole operating dealer in Rebkong, Qinghai Province. No shop was exclusively dedicated to selling SHS equipment, with each shop featuring a variety of electronics, Buddha statues, hygiene products, and apparel among other merchandise. Their primary customers were nomadic herders and dealers rely on word of mouth and repeat customers instead of advertising. Still, customers may ask for specific brands like NIDA or NIMA who advertise on the radio. As many nomads have been using SHSs for more than a decade, customers often do not need information about systems or assistance in their installation and care. Several shops operate on concession agreements with the local PV companies, sometimes through exclusive arrangements. Consequently, dealers remarked that sales have "steadily improved" after the implementation of the REDP.

Improved Economic Competiveness

Before the REDP was implemented, insufficient access to enterprise capital was seen as a key impediment for PV manufacturers and the domestic solar industry.[17] Under the REDP, 28 companies responsible for 90 percent of SHS sales were invited to benefit from the MDSF and TI components. These companies saw significant improvement in various facets of their operations and management. Over the course of the REDP, SHS sales revenues at these companies grew by more than a factor of four; profits grew more than 300 percent; sales in MWp expanded 651 percent; employment doubled; the geographic size of distribution networks doubled; and annual sales grew from about 40,000 units to a peak of 100,000 units. Furthermore, 14 companies invested in new and expanded factories, with factory floor space doubling and office space increasing 50 percent, and 13 companies invested in PV module laminating plants as a way to secure access to supplies and lower costs. National production capacity grew from a meager 10 MWp in 2000 to 2,800 MWp at the close of the REDP in 2007

---

17   World Bank, *Project Performance Assessment Report: People's Republic of China Renewable Energy Development Project* (Wasnington, DC: orld Bank Independent Evaluation Group, 2010).

and 3,850 MWp in 2009,[18] and ten Chinese PV companies are now listed on the New York and London stock exchanges with market capitalization of more than $24 billion.

One company we visited, for instance, told us that they sold "several thousand units" under the REDP and that it was "instrumental" in their business growth. The manager there stated that the "REDP was very successful, it helped a company like mine build capacity and sell more SHSs, research new products, and improve quality." That manager specifically noted that the REDP enabled him to upgrade to better controllers. The owner of another PV manufacturer we visited noted that "the REDP really benefitted our company, it was so great because it gave us funding and support for new technology, new products, standardization, certification, training, and new ways of thinking that have ensured we remain innovative and profitable." A final manager at a third PV manufacturer the research team interviewed mentioned that "without the REDP, many of the present private solar PV companies would have gone out of business ... If the REDP had not been there, the number of existing companies would be drastically lower."

Challenges

Despite these benefits, the REDP also faced (and continues to face) a series of challenges related to credit access, technology improvement, after sales service, grid electrification, and competition with other programs.

Credit Access

Herders and many rural households have seasonal changes in income, making it difficult to afford a SHS when times are bad. When we visited, for example, caterpillar fungus was selling well, a plus, but a 7.9 magnitude earthquake in Sichuan Province killed 70,000 people and disaster recovery efforts rapidly increased the demand for portable forms of energy supply, meaning local SHS prices jumped in some places by 50 to 75 percent for half of 2008. Interestingly, the PMO was originally tasked with investigating, recommending, and supporting mechanisms that would encourage consumer credit access for purchasing SHS. Various intermediaries were to be considered, including consumer banks, rural credit cooperatives, and the PV companies themselves. However, the credit pilot was dropped in September 2005. Had consumer credit access been made available, perhaps the number of beneficiaries in the REDP would have increased.

---

18   L. Junfeng, Y.-h. Wan, J.M. Ohi, Renewable energy development in China: Resource assessment, technology status, and greenhouse gas mitigation potential, *Applied Energy* 56, 381–394.

Technology Improvement

Just as retailers and PV companies stated that end-users were gradually upsizing their SHS units as growing household incomes allow, PV companies also stated that new programs need to focus on larger sized products. Several interviewees identified 3kW and larger, grid-connected systems as the future of their businesses, not smaller SHS units. They also expressed interest in receiving technical assistance to help with product development in new areas like solar pumping systems, solar lanterns for parks, and building-integrated photovoltaics, which paid consultants could provide. Getting to that point requires financial assistance and companies said they would like to see credit facilities that would lend on the scale of RMB 3–5 million with repayment terms stretching beyond five years.

Consequently, assuming the companies in question are indeed credit-worthy, rural credit constraints have impeded both end-users and corporate investments in capital stock. It is worth speculating what the current state of formal credit availability would be for such lenders, had a stronger push for rural financial services been made during the early stages of REDP. This is especially pertinent given the aversion of commercial banks in other countries have had in lending to renewable energy projects or research and development activities, because of asymmetric information about project risk and regulatory uncertainty. Were these channels established earlier, lenders would have had a longer period of time to gain familiarity with the renewable energy industry and increase their lending willingness.

In addition, while the quality of Chinese solar technology has improved, it is far from perfect. One respondent noted that "After the REDP, some manufacturers are still selling bad products, falsifying labels, or presenting second hand units as if they were new." This problem is partly connected to lack of consumer knowledge, education, and awareness, as well as a greater orientation towards exports. One respondent stated that "many customers do not care about buying a certified solar panel, either they do not have enough money to afford it, preferring a cheaper system, or lack information about certificates and don't understand the need for them." During our own interviews it became apparent that many herders underestimate the importance of inverters and controllers, as well as proper battery charging. Given that 80 percent of herders and nomads are illiterate this shouldn't come as a surprise. Another argued that the "best" technology is reserved for exports, while only "low-grade" units are saved for domestic use, creating a "bifurcated market." As this respondent elaborated, "the majority of Chinese manufacturers still focus on large modules, greater than 180 Wp, which constitute less than 10 percent of solar modules sold as SHS in China, and the solar market remains heavily focused on exports, 95 percent of Chinese modules go to North America and Europe."

After Sales Service

Improving the quality and availability of after-sales service at the township and provincial level was another core objective of project developers, but our interviews with end-users indicated that service channels were weak. Damaged or partially operating systems were not uncommon among the households we visited and end-users appeared to possess inadequate knowledge about proper system maintenance. Because retailers in these areas were the primary service centers, operating at the county-level and therefore remote from rural nomads, to repair a damaged SHS incurs transportation costs and time. According to the 2007 survey conducted in Gansu Province and Tibet, 38 percent of respondents cited difficulty finding repair shops as a reason for stopping SHS use.[19]

Grid Electrification

For many rural customers and some herders, SHS provides only a temporary solution en route to higher wattage power sources such as grid-fed electricity. Non-nomadic rural users who gain grid access may no longer need their SHSs. For these users, SHS deployment projects provide a stream of benefits up until that point. As evidenced by the REDP surveys, the broader literature on SHS impacts, and our interviews, SHS provision enables immediate access to lighting and other energy services, such as milk separation, with potential increases in household income. SHSs that have become redundant before the end of their operating life because of grid connectivity have a shorter window of accrued savings from avoided kerosene or candle purchases, decreasing their cost-effectiveness. As one respondent put it, "this year is not a good one for SHS, electricity access is increasing through grid-connected hydro facilities, and herders are also being urged to relocate to permanent houses rather than collapsible tents." Figure 4.6 shows one such new mini-hydro plant near Qinghai, plants that mitigate the desirability and profitability of purchasing a SHS.

Market Saturation

The REDP had to complete with other programs and, perhaps the victim of its own success, market saturation. The Chinese government initiated several renewable energy promotion programs over the last two decades, with support from bilateral and multilateral agencies. The Brightness Rural Electrification Program, for example, originated in 1998 and aimed to bring electricity to 23 million people by 2010, using renewable energy technologies at both the household- and village

---

19   REDP, *2007 Solar Home System End User Investigation Report* (Washington, DC: World Bank, 2007).

Figure 4.6   A 16.5 kW Grid-connected small hydro facility near Rebgong, China

level.[20] The Song Dian Dao Xiang (SDDX), or Township Electrification Program, employed small-hydro, wind, and PV technologies to expand electricity to nearly one million people living in 1,000 townships in Western China.[21] The Song Dian Dao Cun (SDDC) program arose as a second phase of the SDDX, with a goal of electrifying 20,000 villages by 2010. Table 4.4 lists other major renewable energy programs in operation between 1998 and 2010. Manufacturer subsidies under the Dutch Silk Road program accounted for 62 percent of production costs, in comparison to the REDP's 20–25 percent.[22] Companies were selective in their participation and would sometimes forego an existing program for one with more lucrative returns awaiting implementation. This was especially the case for programs that included after-sales support opportunities. As previously mentioned, in areas with available electricity programs also competed with existing energy infrastructure which could produce electricity at lower costs.

These efforts, in tandem, may have saturated the market for SHS. As one respondent put it, "the market is kind of full ... the only demand now is for repairing or replacing units, not selling to new customers, it is much harder to sell SHS in this area." Another remarked that "people are still waiting to be connected to the grid, or for government programs to give them cheaper or even free SHS." One of

20   NREL, *Renewable Energy in China: Brightness Rural Electrification Program* (Golden: U.S. DOE, 2004).

21   H. Liming, A study of China-India cooperation in renewable energy field, *Renewable and Sustainable Energy Reviews* 11 (2007), 1739–1757.

22   REDP 2004.

Table 4.4 Selected PV programs operating in rural China and their estimated duration

Program	1998	1999	2000	2001	2002	2003	2004	2005	2006	2007	2008	2009	2010
NEDO PV Program	■	■	■										
Brightness Program Pilot Project				■	■	■	■	■	■	■	■	■	■
IM Electrify			■	■	■	■	■	■	■				
Silk Road Brightness Program				■	■	■	■	■					
Township Electrification Program (SDDX)						■	■	■					
Renewable Energy Development Project (REDP)							■	■	■	■	■		
KfW Entwicklungsbank / Ministry of Finance Village Electrification										■	■		
GTZ and Department of Commerce, Tibet Autonomous Region									■	■	■	■	■

the shop owners the research team visited lamented that "now SHS accounts for only 20 percent of my annual sales but take up half of my floor space and almost two-thirds my time." Another manager said that "the market for SHS is clearly saturated, opportunities for selling are very limited, the market is full."

Conclusion

The REDP exceeded its targets of installed PV capacity and contributed to renewable energy suppliers manufacturing products in compliance with international standards, enabling them to enter export markets. While some of its two million SHS beneficiaries would have ultimately received SHS units from other programs, or gained grid connections, the project expedited the process in areas where energy poverty had become chronic and widespread. The relative contribution of REDP to China's evolving policy landscape is difficult to measure, but it likely played a significant role in highlighting the merits of renewable energy in combating air pollution, offering energy amenities to households reliant on less modern energy carriers, and supporting the development of the country's growing solar industries.

REDP was highly successful in demonstrating the value of SHS ownership and strengthening fledgling PV manufacturers to improve their business operations and technical capacity through product development, testing, and certification. As evidenced by our interviews with PV companies, they are becoming internationally competitive with exports comprising a growing share of their revenue, made possible by the introduction of quality standards and adherence to international certification criteria. Companies benefited from knock-on effects, not only by transferring technology advancements from the TI component to non-SHS product lines and improving their competitiveness, but also leveraging these gains to win SHS contracts both inside and outside of China. The synergy of simultaneously focusing on industry strengthening, tightening quality standards, incentivizing sales into remote areas through PV sub-grants, and showcasing the value of SHS through roadshows and videos helped make the 'REDP a success. As such, a key takeaway for PV deployment design consists of identifying each stage of the project life-cycle, from product design to after-sales care, and devising components that target the stakeholder's immediate needs. When these disparate factors come together under a single program, the REDP suggests that markets for solar energy not only grow, they sustain themselves to the point of saturation.

Chapter 5

The Rural Electrification Project in Laos

Introduction

Laos, a least developed land-locked country in Southeast Asia, has tremendous potential for hydropower projects because of its access to the tributaries of the Mekong River, generous rainfall, hilly terrain, and a low population density that limits the need for human resettlement along rivers. It also shares a border with countries such as Cambodia, Thailand, and Vietnam, keen to import Laotian electricity to meet their domestic demand. Despite its abundance of hydroelectric resources, however, Laotian energy planners face problems supplying energy services to a large proportion of their population.[1]

For example, the country's low population density and rugged terrain make it difficult and costly to connect everyone to an electricity grid. This gives rise to Laos's energy conundrum: while there is plentiful electricity for export, providing domestic access is difficult. This situation has avoided political criticism because the government has exhibited a strong commitment towards serving domestic customers. As a result of its effort, the government has worked together with the World Bank and other organizations over several decades to bring about access to energy services to poor and rural communities. These seem to have paid off, and Figure 5.1 shows that in 2009, Laos reached an electrification rate of 63 percent, quadrupling from its electrification rate of 16 percent in 1995. The latest of these efforts is the Rural Electrification Project Phase I (REP I), under which the government undertook rural electrification through both grid and off-grid measures. When looking at Figure 5.1, we must emphasize that the REP I has worked in tandem with other national efforts that have in aggregate electrified more than 600,000 homes.

This chapter examines the hydroelectric and SHS components of REP I. Through the program, Laos managed to serve more than 65,000 rural households with hydroelectricity and supply off-grid SHS to almost 17,000 households. Moreover, as Laos is a least developed country (LDC), it faces a unique set of capacity and infrastructure-related barriers unlike high-income economies. Its government institutions and policies are in a nascent stage of development, and it lacks as the capital to finance, and at times even repair, power plants, transmission towers, transmission and distribution lines, substations, and transformers. But it has found ways and means to overcome many of these shortfalls, and the country

---

1   Smits, M. and Bush, S., 2009. A Light Left in the Dark: The Practice and Politics of Pico-hydropower in the Lao PDR. *Energy Policy* 38, 116–127.

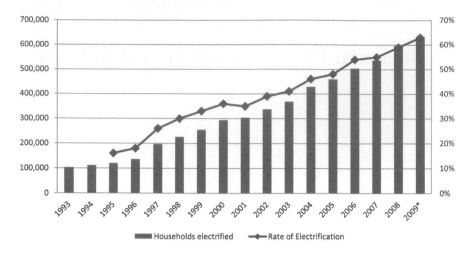

Figure 5.1   National electrification rate in Laos, 1993–2009

offers lessons for how to provide energy access in the face of daunting technical and economic challenges.

Description and Background

Before 2006, the World Bank's assistance came in the form of $97 million of International Development Association (IDA) grants for three separate electrification projects to extend grid-connected electricity to roughly 100,000 households. These projects are listed in Table 5.1. To further enhance access, the REP I started in 2006, when the World Bank provided an IDA grant of $10 million and an additional GEF grant of $3.75 million to the Lao government from 2006 to 2009, later extended to 2010. It was a part of the World Bank's Adaptable Program Loan (APL), which is a long-term loan program with periodic milestones and triggers to transitional loan phases. This allows flexibility and long-term support for what the World Bank calls "lasting sector reforms."[2]

---

2   See Laos Ministry of Mines and Energy 2009. *History of Electrification in Laos.* Vientiane: MME; World Bank 2005. *Information Completion Report on Southern Provinces Rural Electrification Project*; World Bank 2006. *Project Appraisal Document on Rural Electrification Phase I Project*; World Bank 2008. *Project Performance Assessment Report on Southern Provinces Rural Electrification Project*; World Bank 2009. *Project Appraisal Document on Rural Electrification Phase II Project.*

Table 5.1   History of rural electrification programs in Laos

Program	IDA grant ($ m)	Total cost ($ m)	Start date	End date	Electrification targets (initial)	Households electrified
Southern Provinces Electrification (SPE)	26	31	Jun 1987	Dec 1994	n/a	8,354
Provincial Grid Integration (PGI)	36	49	1993	1999	n/a	40,100
Southern Provinces Rural Electrification (SPRE)	35	42	Aug 1998	Dec 2004	50,000 households on-grid, 4,600 off-grid	51,805 and 4,910
Rural Electrification Phase I (REP I)	10 (GEF: 4)	36	Feb 2006	Mar 2010	42,000 households on-grid, 10,000 off-grid	65,706 and 16,692 (above targets)
Rural Electrification Phase II (REP II)	20	36	Jan 2010	Dec 2013	27,700 households on grid, 10,000 off-grid	Ongoing

The objectives of the REP APL Program were set down as twofold: (i) To increase access to electricity of rural households in villages of targeted provinces (seven Southern provinces, as Northern provinces had been electrified by the Asian Development Bank (ADB); and (ii) to achieve sustainability of power sector development. Broader objectives also consisted of poverty reduction and environmental sustainability.

The key stakeholders in the REP I were:

- *The World Bank*, which provided the main funding through an IDA grant, and its various partners such as the Global Environment Facility, Asia Sustainable and Alternative Energy Program, Energy Sector Management Assistance Program, Policy and Human Resources Development Fund, the ADB, the Japanese International Cooperation Agency, Japan Bank for International Cooperation, and the Australian Agency for International Development.
- *Electricité du Laos* (EdL), the state-owned utility, which implemented hydroelectric grid expansion. It employs more than 3,000 people, and operates the country's electricity generation, transmission and distribution assets. It generates and distributes power under direction from the Department of Electricity, housed within the Ministry of Energy and Mines.
- *The Laos Ministry of Energy and Mines* (MEM) and the various bodies it involved in implementing off-grid electrification such as the French consultancy Innovation Energie Developpement (IED), village off-grid promotion and support (VOPS) offices, provincial electricity

supply companies (PESCOs), village electricity managers (VEMs), and policymakers at the provincial departments of energy and mines.

Electrification in the REP I was carried out by extending both grid access to small-scale hydroelectricity and off-grid access to SHSs, described in more detail below. This is a reflection of the World Bank's recognition that off-grid extension is a viable alternative to grid expansion. A final element of the REP I involved capacity building. The World Bank provided funding for assessments of national energy policy and any requisite regulatory modifications needed to ensure the project was implemented smoothly. Training was also offered to members of PESCOs so that they could better understand, and deliver, SHS units to rural communities. Planners possessed an appreciation that SHSs might not be the only suitable off-grid technology for Laos, and provided a small amount of support to develop delivery models and studies of microhydro potential, including a feasibility study for 13 microhydro project sites. An off-grid operations manual was also printed and published for PECOS and VEMs.[3]

Lastly, electrification efforts went hand-in-hand with energy efficiency targets. Electrification during the earlier SPRE project had "overstretched" the transmission network because the existing infrastructure of medium- and low-voltage lines were being used to cover long distribution networks in order to meet targeted electrification numbers. During REP I, this previous oversight was recognized and emphasis was placed on loss reduction and upfront optimization in grid extension design. A "Master Plan for Distribution Loss Reduction" was developed, which included investing in state-of-the-art software and hardware, project evaluation methodologies for reducing technical losses, computerized billing and accounting systems, and field measurements for non-technical losses. As a result, system losses dropped from 20 percent in 2005 to 13 percent in 2009.

Benefits

Overall, the REP I project was highly successful in achieving high electrification rates for both grid-connected small hydroelectricity and off-grid SHS, meeting all of its targets ahead of schedule.

Grid Extension Component

For the hydroelectric grid-extension component, 36,700 households were electrified by mid-2009, 49,397 by the end of 2010, and 65,706 by the middle of 2011, far above the revised target of 42,000 households; 570 villages were

---

3   Fraser, J. and Tang, J. 2011. *Lao PDR: Implementation Support for the Rural Electrification Program (REP)*. Vientiene: World Bank.

connected by 2010 and 671 were electrified as of mid-2011, well above the target of 540, though the REP did receive additional funding from international donors.[4]

This grid extension was carried out by EdL, who built small-scale hydroelectric dams and then connected them to 22 kV, 12.7 kV, and 0.4 kV lines, transformers, and single wire earth return (SWER) systems. Villages for electrification were chosen with preference given to those closer to existing roads, larger in size, and already engaged in economic and social activities. Efforts were also put into transmission and distribution loss reduction through better systemic planning and preparation for REP I, cutting distribution losses almost in half from 2005 to 2009. In addition, a demand-side management (DSM) division was established within EdL to implement DSM and energy efficiency (EE) programs. Construction and installation teams were set up in EDL's five branch offices across various provinces, and these teams reported to the headquarters.

There were several factors that led to the scheme's success. Perhaps the most significant was its flexibility in determining whether a community should be electrified by a hydroelectric grid or served with SHSs. Regarding this choice between grid expansion and off-grid electrification using renewable energy, some interviewees argued that the grid may have been stretched to areas that were beyond economic and technical feasibility. According to one respondent, though the marginal cost of bringing in grid electricity to a household in Laos will depend on the distance from the grid, with an average cost of $350 to $400 per connection, in extreme cases where a substation needs to be built to reach a remote small-sized village, the mean amount can be as high as $1,000 per connection. These figures are also supported by older World Bank estimates where the general costs of grid-based rural electrification extensions were calculated to range from $230 to $1,800 per connection, with a median cost of about $520 per connection.[5] In contrast, an off-grid 50Wp SHS needs an average subsidy of only $150 to $200 per unit. Therefore grid extension into remote parts of Laos is not only expensive (depending on distance and number of households), but could arguably also be wasteful in comparison to SHSs.

However, the other side of the argument is that the electricity generated through SHSs is not as effective as through grid connections. According to the IFC, "Experience has also demonstrated that people are looking for a constant supply of electricity provided by grid connection. … solar PV simply cannot provide equivalent services to the grid, and it is also not the only technology available for

---

4　Some of these households were likely electrified as the result of REP Phase 2, which started in late 2010, though precise figures are difficult to determine given that the phases overlap. See Fraser, J. and Tang, J. 2011 *Lao PDR: Implementation Support for the Rural Electrification Program (REP)*. Vientiene: World Bank; Fraser, J. 2011 *Implementation Status & Results, Lao People's Democratic Republic Rural Electrification Phase I Project of the Rural Electrification (APL) Program (P075531)*. Report No: ISR3174.

5　Mason, M., 1990. Rural Electrification: A Review of World Bank and USAID–Financed Projects. *World Bank Background Paper*, Washington, DC.

addressing rural electrification demand."[6] During our interviews at Mai village, which was provided with off-grid electrification using solar panels, some of the respondents mentioned that they would prefer to be connected through the grid so that they could, for example, have access to refrigeration to store fish. This raises questions about the appropriateness of SHSs for the needs of the people, their ability to pay for the technology they desire, and whether intermediate solutions such as village-level micro-grids may offer a "middle of the road" option between grid expansion and household level off-grid electrification.

Rather than promote grid connection or SHS entirely as a "one-size fits all" approach, the REP instead extended the grid to areas that were cost competitive with SHS, and then relied on SHS for areas where grid extension was too expensive. During the previous project SPRE, the PESCOs (with the help of the DoE and the EdL) chose villages for off-grid electricity implementation by identifying those that were not likely to be linked to the grid for at least ten years. If the grid came earlier than expected, then the off-grid system would be moved into a more remote village. This process was not very efficient, as it involved duplication of electrification efforts.

Hence, the REP featured a renewable electricity master plan. A GIS (geographic information system) based database was created to help planners determine which option—"grid" or "solar"—would work best for each targeted community. This GIS database detailed a map showing both on- and off-grid areas combined with village socio-economic status. This assisted government planners and REP managers to better comprehend which options to promote where.

Other factors played a prominent role in determining the success of the project: firstly, how EdL would viably carry out its social objectives of providing grid electricity to the maximum number of people at the most affordable rates, but at the same time improve its financial performance; and secondly, how EdL would expand the grid without overstretching it.

The EdL was tasked with expanding the grid to new households, but at the same time keeping tariff rates affordable. Tariffs for residential consumers were therefore cross-subsidized against other consumer cohorts. The 2008 average tariffs in Figure 5.2 show that the residential sector tariff was one of the lowest, and below the average tariff of 542 Kip/kWh (or a little more than 6 ¢/kWh). This low tariff was designed with the social objective of providing as many households as possible with electricity. Furthermore, Figure 5.3 shows that the average tariff in Laos was still in fact one of the lowest in the ASEAN region despite its cross subsidization of grid extension efforts.

As an extension to the principle of affordable tariffs, a special component called the Power to the Poor (P2P) project was also launched during the REP I. P2P aimed to include those households that did not sign up for grid connection even though the grid was passing through their village, due to the unaffordability of upfront

---

6   IFC 2007. *Selling Solar: Lessons from More than a Decade of IFC's Experience.* International Finance Corporation, Washington, DC.

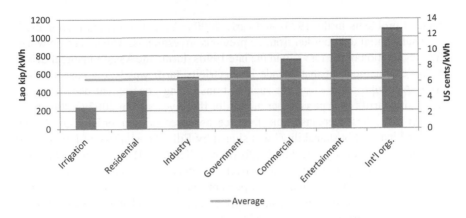

Figure 5.2   Average sector electricity tariffs in Laos, 2008

connection charges. The program provided these households with a low voltage 3/9 ampere meter, and an interest-free loan of up to 700,000 Kip (~$80) for upfront costs. This was to be repaid in small monthly installments over a maximum of 3 years. The program targeted women, as it was found that households that refused electricity connections were typically headed by females. The scheme, which ran from late 2008 to January 2009, connected more than 500 households in less than six months. It raised connection rates in these villages collectively from 78 percent to 95 percent. As a result of this success, is the P2P project is planned to be scaled up in the second phase of the REP.

However, at the same time that EdL was tasked with these objectives, it was also being pushed on a path of commercial viability. EdL had been plagued by serious

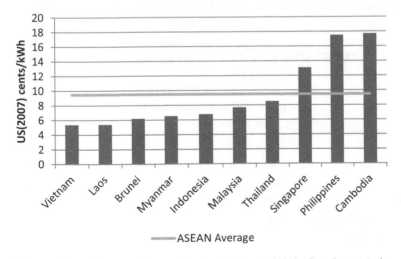

Figure 5.3   Average domestic electricity tariffs in Southeast Asia

financial problems in the 1990s and early 2000s. As a state owned monopoly, it had a substantial debt burden from its previous investments, which became worse after devaluation of the Lao Kip in the late 1990s during the Asian Financial Crisis. Moreover, for much of that decade its electricity tariff was below cost-recovery levels. Although some measures such as transferring government debt to equity and raising tariffs by 2 percent per month for three years from 2002 to 2005 were taken in 2001, these were insufficient. Hence, as a precondition for REP, an "Action Plan for Financial Sustainability" was signed by EdL, the Ministry of Finance and the Ministry of Industry and Handicrafts (now called MEM) in November of 2005. This led EdL on a path to financial recovery in several ways: tariffs were raised to cover operating costs and achieve a 4 percent real return on fixed assets by 2011. They were also linked to the consumer price index and exchange rates in order to ensure that real tariffs rose by acceptable rates.

The action plan pressured the government to pay off EdL's electricity bills. In the past, bills payable by the government had been offset against other financial favors to EdL, but the World Bank pushed for improved governance and accountability. EdL was also allowed to retain dividends from export projects, such as the Theun Hinboun hydropower project (built in 1998 at a capacity 220 MW) in which EdL has a 60 percent share. This measure was provided as an operating subsidy by the government on a declining basis, i.e., it was phased out after 2011 once tariffs were raised. Lastly, EdL was given a capital subsidy by transferring 80 percent of the IDA fund on a grant basis for the on-grid program, and 100 percent of the GEF funds on a grant basis for capital costs incurred in the off-grid program. As a result, EdL has steadily improved in performance and has even financed 25 percent of the REP I grid extension program on its own (see Table 5.2).

Hydroelectric grid extension has clearly benefitted local communities. In Nongsa village, visited by the authors, which had been provided with grid access in 2009, households were relatively wealthier than in off-grid areas. In addition

Table 5.2   Financing plan for REP I

EdL component	Source ($ million)					
	IDA	NORAD Co-financing	GEF	EdL	Consumer contributions	Sub-total
Grid extension	5.56	10		6.61	4.23	26.4
Loss reduction	1.00			1.00		2.00
IT system	0.80					0.80
Tariff reform	0.05					0.05
Safeguards capacity-building	0.14					0.14
DSM and energy efficiency			0.75			0.75
Sub-total	7.55	10	0.75	7.61	4.23	30.14

to TVs and lamps, most households used grid electricity for refrigerators. Their key sources of income were managing rice plantations and raising livestock. One house we visited doubled up as a restaurant to serve children from a local school nearby. Community members told us that grid electricity has changed their daily life in several ways, by providing extended daylight hours, enabling access to refrigeration and electric ironing, facilitating the sale of perishable food items in their restaurant, allowing electric water pumping, running electric rice mills, and making available entertainment through TV and radio.

Off-Grid SHS Component

For off-grid electrification, SHSs, consisting of solar panels (sizes varying from 20 Wp to 50 Wp) along with batteries, load controllers and high efficiency lamps, were installed in more than 16,000 households across 488 villages in 16 provinces as of mid-2011. The REP I followed an energy service company (ESCO) based leasing model, i.e., they leased the equipment to households for a monthly fee through companies.

But neither the government department nor EdL had the institutional capacity to support installation and implementation of off-grid electrification. Hence it was decided that an external contractor should be hired to support this work, at least for the first phase of three years, during which time the government could build its capacity. IED was awarded the implementation contract and set up a management structure involving VOPs, PESCOs, and VEMs, as illustrated schematically in Figure 5.4.

As this figure shows, the scheme established VOPS offices at the upper level to manage PESCOs, which run operations at the provincial levels. PESCOs are local private companies responsible for planning, installation, operations and maintenance of the off-grid systems. They are paid $2 for each household that signs up, $1 for installation, and 20–35 percent of the monthly payments collected from the units in operation. PESCO managers are recruited by VOPS. PESCOs

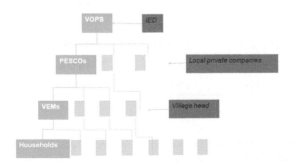

Figure 5.4   Organization of the off-grid component of REP I in Laos

work with a VEM, who is in effect the head of the village and makes decisions for the village and its residents. VEMs also keep 20–35 percent of the monthly payments as fees for their services. The remainder of the monthly payments is put into a rural electrification fund (REF). This fund is supervised by VOPs and used to pay for additional incentives to PESCOs, VEMs, and for supplies of spare parts and batteries.

In addition to PESCOs and VEMs, other key actors include Provincial Departments of Energy and Mines (PDEMs) who regulate PESCOs and inspect the quality of installations, and a Village Electricity Advisory Committee (VEAC) which oversees each program at the village level. Procurement of the solar panels was mainly through Sunlabob, a private entrepreneur who coordinated the purchases of all SHS components (panels, wires, controllers, battery, switches, inverters, and so on) from various suppliers across Asia, and then assembled them into a complete system.

Monthly lease terms for SHS were set keeping the principle of affordability in mind, to ensure satisfactory adoption rates. Affordable and accessible financing is "a major consideration in the design of any PV program due to the high first costs of solar home systems,"[7] and "deeply ingrained as a first principle into the World Bank's ethos," according to one respondent. Hence, in the REP I, one basic criterion for taking a SHS into a village was that at least half the households there had to express an interest in purchasing it. Tariffs were set by spreading the capital and installation costs over either a five-year or ten-year period—both term choices were made available to households, VEMs, and VEACs (see Table 5.3). The positive side of this was the high adoption and uptake of the technology. Uptake rates finally achieved in villages where SHS was installed were, on average, 80 percent, and as of mid-2011 off-grid renewable sources of energy reportedly met 21 percent of national rural household energy demand.[8]

Table 5.3   **SHS installation and monthly payments (in Lao Kip)**

	Upfront Installation Cost	Direct Purchase	Monthly Lease for 5 Years	Monthly Lease for 10 Years
20Wp	160,000	1,560,000	26,000	13,000
30Wp	190,000	2,160,000	36,000	18,000
40Wp	220,000	2,880,000	48,000	24,000
50Wp	250,000	3,600,000	60,000	30,000

---

7   Cabraal, A., 1996. *Best Practices for Photovoltaic Household Electrification Programs: Lessons from Experiences in Selected Countries.* World Bank Technical Paper number 324, Asia Technical Department Series.

8   Fraser 2011.

Like its grid counterpart, SHS have invariably improved the lives of communities adopting them. In the off-grid Mai village, which we visited, 72 of the 101 households had SHS and a strong majority were "pleased" and "very happy" with them. Based on conversations with the villagers and a visit to the VEM's house, we learned that these households typically linked their SHS to a TV and two lamps. Many had replaced the original battery with cheaper car batteries. With a 50 Wp solar panel, the most commonly used, they were able to run the TV for about an hour and the lamps for about 2–3 hours every day. When asked about the satisfaction level with the current system, the VEM mentioned that many households would prefer to have the grid as it would allow them to use electricity for other purposes such as powering rice mills or refrigerators to store caught fish. When asked how electricity had changed their lives, extended daylight hours were cited as one of the most significant benefits, which helped them to do basket weaving at night or allow children to study after sunset.

Challenges

However, the REP also confronted challenges related to affordability and technology size, financial solvency, its effect on the domestic renewable energy market, and institutional capacity.

Firstly, the economic feasibility of intermediary institutions such as VEMs and PESCOs in Laos is questionable. The PESCOs must earn their revenue by retaining a proportion of the end-user tariff, and both the percent of their commission and the end-user tariff are fixed. As a result, PESCOs are not free to set the lease terms of SHSs. Predictably, during the project some PESCOs complained that REP-I fee levels were not enough to sustain their operations to cover the costs of regular fee collections, let alone to cover the costs of repairs and maintenance of installed SHS. For example, follow-up assessments by the World Bank have revealed that the expenditures incurred from installing small equipment like a 20 Wp SHS incur more costs than the benefits they provide; that is, those systems result in a "negative inflow" of revenue.[9]

Because they are so decentralized and democratic, performance also varies greatly between PESCOs and VEMs. The World Bank has recently cautioned that "several PESCOs" have "very poor performance," are not placing funds in to the REF account as promised, and are accruing substantial debts.[10] For example, PESCOs reported 150 million Kip of debt as of June 2010 but this has grown to 358 million Kip in December 2010 and 471 million Kip as of March 2011.[11] This led one interviewee to speculate that "many PESCOs may soon be bankrupt."

---

9 Julia Fraser and Jie Tang, *Lao PDR: Implementation Support for the Rural Electrification Program* (REP) (Vientiene: World Bank, May, 2011).

10 Fraser and Tang 2011.

11 Fraser and Tang 2011.

Secondly, the REP I has had some negative influence on the domestic renewable energy market. The SHS provided by the REP I competed directly with Sunlabob's solar rental business. Sunlabob is a private sector company that rents out solar energy systems to rural households. Before the REP I, Sunlabob had its own model for SHS distribution and operated through 36 franchises in the country. These franchises, although employed by Sunlabob, worked independently in doing their own marketing and sales. They built a network of distributors by going through a village union such as a women's cooperative. These village unions were also made responsible for collection of the rent for the system, and for their services, they were allowed to keep a portion of the rent as commission. The rent for a 20 Wp solar energy system was about $3.50 (~30,000 Lao Kip) per month, including maintenance and servicing. The REP project, subsidized from the World Bank and other international actors, came in with a similar model of renting out solar home systems, but at less than half the cost of Sunlabob's rentals. As a result, Sunlabob's business model was "rendered functionally uncompetitive" by the REP I project.

Third, building institutional capacity has faced a series of obstacles. During the predecessor of REP I, the SPRE project, it became clear that neither EdL nor the government had the resources and capacity to handle off-grid rural electrification into remote villages. Hence the VOPS implementation work was contracted to IED who was employed for a fee for 3 years between 2006 and 2008. This turned out to be a successful strategy in terms of bringing a high level of efficiency into the operations and achieving the quantity of installations within the REP's ambitious timeframe.

However, the aim was for the MEM to build its own capacity by the end of 2008. As this did not happen, IED's contract was extended by a year till the end of 2009. However, even by that time, the MEM reported that it was "not ready" to handle the operations by themselves. Interviewees attributed this lack of capacity to "lack of funds" and "low availability of the requisite talent—particularly of personnel who are conversant in English and can deal with project implementation with external agencies." Apparently English language training programs provided by the World Bank did not achieve their desired results. As a result, another external consultant has been contracted to assume the VOPS role for 2010 and 2011. However, this dependence on aid-sponsored outsourcing is clearly hampering domestic institution and capacity-building. If the laudable trend towards rural and off-grid electrification in Laos is to continue, perhaps more rigorous measures and incentives will be needed to strengthen domestic institutions.

Conclusion

The REP I was extremely successful in terms of reaching its targeted number of households through both hydroelectric grid expansion and off-grid SHS electrification, serving to underline the government's strong commitment to increase access to basic energy services for rural and poor communities. Rather than promoting a "one-size fits all" solution, however, the project's laudable innovation was the creation of a centralized geographic database mapping potential grid and off-grid areas along with socio-economic factors, enabling planners and managers to make better calculations and be more flexible regarding the suitability of grid extension versus off-grid technology applications in different parts of Laos. The off-grid SHS component was also novel in the way it decentralized authority to provincial and village officials and companies, placing solar energy "directly in the hands of those that will be using it," as one official put it.

Nevertheless, the hydroelectric grid-extension component was clearly the backbone of the project and a policy priority for the government, successfully reaching 65,706 households by the middle of 2011, far above the revised target of 42,000 households. The scheme was able to maximize the number of households that could be connected at affordable tariffs, without compromising the financial performance of EdL or overstretching the capacity of the grid. To further demonstrate the compatibility of social objectives with a profit-oriented endeavor, a special interest-free loan under the P2P component was offered to households that were unable afford the upfront connection charges, with positive results. The implication is that for Laos, at least, grid extension, commercial profitability, and the eradication of energy poverty need not conflict with each other, and can go hand-in-hand.

Chapter 6

The Rural Electricity Access Project in Mongolia

Introduction

Like many other developing countries, Mongolia confronts high rates of poverty, low population density, and low per capita electricity use. With a per capita income of less than $500 per year, Mongolia is what the World Bank calls "one of the least developed countries in Asia."[1] The transition from a centrally-planned economy to a market economy after the collapse of the Soviet Union placed remarkable stress on its national energy sector. The United Nations Development Program (UNDP) reports that "a substantial portion of the population in Mongolia still lacks access to electricity despite an expansion in the country's energy infrastructure."[2] The United Nations Economic and Social Commission for Asia and the Pacific (UNESCAP) estimates that one-third of the population currently lacks access to electricity and almost half (43 percent) lack access to central heating.[3] The country's large geographic size makes transmitting and distributing electricity difficult, a feat compounded by old and inefficient Russian generators that frequently break down and are completely dependent on fossil fuels. As the World Bank also noted, "weak rural electrification planning" and an "absence of integrated rural development planning" had culminated in a renewable energy market "slow to develop" by 2005.[4]

One novel way of overcoming these obstacles and providing Mongolia's nomadic herders, a highly dispersed and constantly relocating population, with energy services is small-scale solar home systems (SHS) and wind turbine systems (WTS). This chapter explores the history, benefits, and challenges facing the

---

1 World Bank 2006a. *Project Brief on a Proposed Grant in the Amount of USD 3 Million Equivalent and Proposed Grant from the Global Environment Facility in the Amount of USD 3.5 million to the Republic of Mongolia for a Renewable Energy and Rural Electricity Access Project*. Washington, DC: World Bank Energy Sector Unit, Infrastructure Department, East Asia and Pacific Region.

2 United Nations Development Program 2008. *East Asia: Mongolia Energy-Efficient Straw-Bale Housing*. Bangkok, UNDP Regional Center.

3 United Nations Economic and Social Commission for Asia and the Pacific 2004. Mongolia's Sustainable Energy Sector Development Strategy. *Presentation to the Concluding Regional Workshop on Strategic Planning and Management of Natural Resources*, Bangkok, Thailand.

4 World Bank 2006a, 22–23.

Renewable Energy and Rural Electricity Access Project (REAP) in Mongolia, an internationally sponsored $23 million program that has thus far distributed more than 40,000 SHS and small-scale WTS to these herders. Estimated at a population of 170,000, Mongolian herders frequently live off-grid in *gers*, collapsible tents that can sleep about four people and can be easily dismantled, and they move according to the weather and grazing conditions.[5] REAP was designed to enable nomadic herders to rely less on coal and fuelwood, and to improve their standard of living by making lighting, refrigeration, communication, television, radio, and cooking more accessible.

An evaluation of REAP, and an exploration into how nomadic herders use renewable energy in Mongolia, is salient for three reasons.

First, the chapter is first and foremost an attempt to inform Mongolian policy. A dearth of academic scholarship has so far explored rural energy use or even energy policy in Mongolia. Assessments in the region often drift towards China (where rapid industrialization has prompted an epic quest to acquire energy fuels and develop energy technologies) or even Kazakhstan (which has plentiful reserves of natural gas and uranium). Yet Mongolia is at a unique crossroads in terms of its energy policy, seeing the emergence of a nascent renewable energy market alongside favoritism towards large-scale fossil-fueled infrastructure. It has an immense amount of renewable resource potential, but the country needs robust and effective public policies if such resources are to be developed. The country possesses solar insolation of 4.5 kWh/m² on a daily average and 1,400 kWh/m² on annual average basis.[6] Put another way, Mongolia averages 270 to 300 sunny days per year with average annual daylight ranging from 2,250 to 3,300 hours. In terms of wind energy, Mongolia's 21 provinces have an astounding 370,000 MW of wind potential[7], enough to provide 548 times the country's current installed electricity capacity, and the Gobi desert and plain zones are renowned for holding a technical potential of 800 billion kWh/year of wind electricity. This chapter has value for informing these policymakers about how they can further expand access to energy services without relying on coal and other big power plants. The government has pledged to provide "reliable access for all" by 2025 and in 2007 passed a renewable energy law, yet whether it continues to invest in conventional or renewable forms of energy supply remains to be seen.

---

5   World Bank 2006b. *Project Information Document Appraisal Stage for the Renewable Energy for Rural Access Project (REAP)*. Washington, DC: World Bank Report Number AB2534.

6   GEF 2006a. *Request for CEO Endorsement GEFSEC PROJECT ID: 2947*. Washington, DC: Global Environment Facility; GEF 2006b. *Project Executive Summary GEF Council Work Program Submission Agency's Project ID: P084766*. Washington, DC: Global Environment Facility.

7   Government of Mongolia 2005. *National Program for Renewable Energy: 2005– 2020*. Unofficial Translation.

Second, REAP demonstrates the environmental benefits of SHS and WTS, especially when they fully or partially displace less efficient and more polluting diesel- or coal-fired devices for cooking, lighting, and heating. Low temperatures throughout much of the year in Mongolia result in significant energy demand for basic heating and cooking services. An urban *ger* will consume a staggering five tons of coal and 1.5 tons of fuelwood *per year*, or more than 20 kilograms of coal per day during the nine winter months. *Gers* with SHS and WTS, however, consume much less conventional fuel, especially when large enough systems are purchased to enable electrical cooking and heating.

Third, REAP had to overcome a variety of external political and cultural factors to succeed. Previous attempts at creating a viable private sector for small-scale renewable energy technologies suffered poor performance in the late 1990s, with small enterprises that were "weak" and prevented by "several market failures" from operating effectively.[8] With support from the Japanese and Chinese governments who donated systems, the Mongolian government initiated the 100,000 Solar *Gers* Program. It was launched in 2001 but "stalled" due to heavy reliance on government disbursements, low private retailer participation, unsatisfactory performance, and customer trepidation over the quality of solar systems.[9] Understanding how REAP overcame these barriers offers insight for other programs attempting to meet the energy needs of rural users.

Description and Background

Before explaining the intricacies of REAP and presenting our findings, it is useful to briefly explore the history of the energy sector in Mongolia, illustrate the technology and capacity of the existing electricity sector, and highlight some key themes regarding rural energy use.

As a country Mongolia is comprised of 21 provinces, called *aimags*, within which there are 329 *soums*, the equivalent of a prefecture. Some herders live in *gers*, collapsible and transportable housing that can be moved virtually anywhere, while others live in permanent *soum centers*, community hubs that include hospitals, schools, and banks to serve the herders. Mongolia's harsh climate and low population density play a defining role in the Mongolian energy sector, which make generating and selling electricity over great distances to *soums* difficult. Extremely cold winters and higher elevations make demand for energy much greater in the winter than the summer.[10]

---

8   GEF 2006b.
9   GEF 2006b.
10   Mehta, A.H., Rao S., and Terway A. 2007. Power Sector Reform in Central Asia: Observations on the Diverse Experiences of Some Formerly Soviet Republics and Mongolia. *Journal of Cleaner Production* 15, 218–234.

The Mongolian energy sector changed dramatically after the fall of the Soviet Union in the early 1990s. Electric utilities and vertically integrated energy companies were given an inordinate number of tasks, not all of them consistent. Among these were generating affordable electricity to commercial firms and residences, earning revenue for the state, and achieving self-sufficiency in energy production. While the utilities struggled to meet these "diverse and conflicting" tasks, inadequate metering, electricity tariffs set below financial recovery levels, corruption, and lack of transparency resulted in the overuse of energy and poor estimations of future energy demand.[11] Such distortions were managed under Soviet rule by large flows of financial aid from Moscow, but that assistance stopped completely after the collapse of the Berlin Wall in 1989.

In 2001 regulators restructured the Mongolian electricity grid, breaking it into 18 independently and publicly owned distribution, transmission, and generation companies. Mongolia's Energy Regulatory Agency now sets electricity tariffs, and the main electricity grid consists of three regional interconnected systems—the Central, Western, and Eastern systems—as well as a number of isolated grids. Of these three, the Central system serving Ulaanbaatar and its surrounding area is the largest, representing 91 percent of installed capacity and 96 percent of all electricity generated.[12] The Central system consists of five coal-fired power plants constituting about 600 MW of capacity generating 3,594 GWh of power in 2009 (See Table 6.1). The Western system mostly imports electricity from the Russian Federation, the Eastern one consists of a single coal-fired combined heat and power facility.

Table 6.1   Basic statistics for the Mongolian electricity sector, 2007

Installed Capacity	674.3 MW
Electricity Generation (gross output)	3,594 GWh
Electricity Imports	130 GWh
Transmission and Distribution Losses	17.4%

The numbers of independent or isolated grids is much greater, and serve four *aimag* centers and 182 *soum* centers through distributed generation, mostly diesel generators. Much of the electricity supply at these centers is intermittent, lasting at best four to six hours a day, and a majority of centers have no metering at

---

11   Mehta, A.H., Rao S., and Terway A. 2007. Power Sector Reform in Central Asia: Observations on the Diverse Experiences of Some Formerly Soviet Republics and Mongolia. *Journal of Cleaner Production* 15, 218–234.

12   Economic Consulting Associates 2009. *Mongolia: Power Sector Development and South Gobi Development*. London: Report Submitted to the World Bank Group.

the household level, instead relying on a single meter for the entire community.[13] The herder population typically resides in rural areas beyond the reach of both the centralized electricity grids and the isolated *aimag* and *soum* grids. Table 6.2 shows that in 2005 almost 200,000 households still lacked access to electricity, including about one-third of all *gers*.

Table 6.2 **Total and percentage of Mongolian households without electricity**

	Total	Houses without Electricity	*Gers* without Electricity
Ulaanbaatar	3,643	938	2,705
Aimag Centers	9,881	1,524	8,357
Villages	2,478	830	1,648
Soum Centers	16,630	5,281	11,349
Outer Rural Households	144,552	11,711	132,841
Total	177,184	20,284	156,900
%		4%	29%

To promote electrification and renewable energy, the government approved a Sustainable Energy Sector Development Strategy Plan (2002–2010) in 2001, which stipulated a government commitment to expanding access to energy services for herders, developing diesel-renewable energy hybrid systems, and reforming *soum* electricity markets to make them more competitive and profitable. Parliament approved a National Renewable Energy Program in 2005 which set a national renewable portfolio standard with a target of 3 to 5 percent of renewable energy supply by 2010 and 20 to 25 percent by 2020. Parliament also passed the Renewable Energy Law of Mongolia in 2008, which included a German-style feed-in tariff for solar and wind energy which passes on premium costs for renewable electricity among all rate-payers.[14]

Despite these efforts, however, some 170,000 herders are nomadic and the government has already determined that connecting the 182 isolated *soum centers*

---

13   World Bank 2006a.

14   Radii, G. 2008. Renewable Energy Resources and the Utilization in Mongolia, presentation to the 4th Annual Meeting of CAREC Electricity Regulators Forum, Karven Issyk-Kul, Kyrgyz Republic, September 15–19. Available at: http://www.adb.org/ Documents/Events/2008/4th-CAREC-Electricity-Regulators/CERF-MON-Renewable-Energy.pdf.

to the grid would be "cost prohibitive".[15] These *soum* centers instead have relied on small diesel generators with high system losses and relatively high generation costs. *Soum centers* usually do not operate their generators during the summer months to keep fuel costs low, and in the winter months only one-third of *soums* operate generators continuously.[16] Left outside the grid, these families are limited to agricultural and livestock production with "few opportunities for non-farm employment or other value-adding economic activities."[17] Trying to engage the private sector to provide them with electricity is problematic and compounded by the lack of information about daily electricity consumption patterns by rate class or end-use and almost a complete lack of metering.[18]

As one interview respondent put it simply, "the central goal of REAP was to increase energy access for off grid rural areas by delivering SHS and small-scale wind turbines to herders and nomads." As Table 6.3 summarizes below, REAP provided $23 million in financing to expand electricity access from 2007 to 2011. Key objectives were to improve the affordability of electricity services among herders and in off-grid *soum centers*, and to "remove barriers to the development and use of renewable energy technologies in grid and off-grid connected systems." The project was divided into three components aimed at herders, *soum centers*, and national capacity building.

Key stakeholders included the National Renewable Energy Center (NREC), formerly the Renewable Energy Corporation of Mongolia, which served as the implementing agency. NREC oversaw contracting, procurement, monitoring, and reporting, and also implemented the capacity-building component. The Mongolian Ministry of Fuel and Energy (MOFE) served as project coordinator. A Project Implementation Unit established by MOFE, staffed with seven employees, was charged with implementing the other two components. The entire project was managed by a Steering Committee consisting of key staff from Ministry of Finance, Ministry of Environment and Resources, and MOFE.

As the following three subsections elaborate, the Herder Electrification Component focused on establishing a rural retail network of private SHSs and WTS with the goal of distributing 50,000 systems to herders over five years. The *Soum center* Component developed the institutional and technical capacity of rural electricity suppliers. The National Capacity Building component developed a regulatory framework for grid-connected renewable energy systems. Table 6.3 outlines the differing objectives, activities, deliverables, and impacts for each of the components.

---

15   GEF 2006a.
16   World Bank 2006b.
17   GEF 2006b, 44.
18   World Bank 2006a.

Table 6.3   Summary of REAP components and objectives

Component	Budget $m	Objective	Activities	Deliverables	Impacts
Herder Electrification	11.60	Strengthen solar home system supply and small-scale wind chains	Train certified suppliers and installers. Offer seminars for *soum*-based technicians; improve testing and standardization. Provide business development loans to certified suppliers so they can increase sales, conduct field visits, diversify the market, and overcome bottlenecks.	Training materials. Seminar evaluations. Reports on service mechanisms for *soum* centers. Analysis of SHS supply chain before and after project.	Greater numbers of certified suppliers. Increased SHS/wind turbine sales. Reduced SHS/wind turbine costs. Improved after-sales service at *soum* centers. Improved SHS/wind turbine quality (longer lifetime).
Soum Center Electricity Services	10.09	Give technical assistance for small-scale grid suppliers.	Analyze needs of *soum* utilities. Understand barriers to grid electrification of rural areas. Offer business development loans and access to financing	Training manuals for *soum* utilities. Reports on electricity supply for *soums* and villages.	Utility service levels increased. Number of middle-sized enterprises now using electricity.
National Capacity Building	1.31	Provide training, monitoring, evaluation, and dissemination.	Integrate monitoring and evaluation activities within REAP. Disseminate lessons learned widely. Host workshops and focus groups.	Workshops hosted Publications drafted.	Feedback from monitoring and evaluation exercises.

Herder's Electricity Access

The Herder's Electricity Access component was intended to develop a legal and institutional framework for rapid deployment of 50,000 SHS and WTS for nomadic herders as well as a battery management program. This component administered subsidies towards first time buyers of SHSs and wind turbine systems, offering to cover 50 percent of the cost of a 20 Wp system (~$80 subsidy) and 40 percent of a 50 Wp system (~$160 subsidy). Private dealers worked directly with the MOFE to receive the subsidy from a "REAP Subsidy Account." Funds were transferred once a purchase of a SHS or WTS by a herder was verified. REAP also established technical standards and procedures for testing the quality of SHS and WTS devices and mandated that only qualified systems could receive the subsidy. The first phase, from 2007 to 2009, was geared towards establishing a rural network and a second phase from 2010 to 2011 was aimed at scaling up that network.

The Herder component also provided marketing support to companies to help them enhance sales and commercialize SHS and WTS technologies. Funds enabled public awareness and consumer engagement through the preparation of sales catalogs, the display of sample systems, the provision of after-sales service, and the establishment of warranties. Figure 6.1 shows one of these display systems outside of a retailer in Ulaanbaatar. Surveys were also conducted to determine potential markets for SHS and WTS as well as to measure consumer acceptance of systems after they had been purchased. The component aggregated the purchase of SHS parts that were imported primarily from China, with the aim of achieving better prices due to bulk purchases. Retailers thus had the option of buying their

Figure 6.1   A four-panel SHS display unit sits outside of the Yahap retailer in Ulaanbaatar, Mongolia

parts from the program office at better prices than if they imported directly on their own.

Soum Center Electricity Service Improvement

The *soum* center component attempted to build capacity within *soum* electric utilities, rehabilitate their transmission and distribution networks, and investigate the feasibility of larger renewable energy micro-grids. Part of this component involved engaging the private sector to participate and invest in *soum* utilities, another part on designing a regulatory framework conducive to more economically sound tariff setting, metering, billing, and "revenue management." Yet another part of the component included the rehabilitation of 30 existing micro-grids to reduce technical losses and the conversion to renewable and hybrid renewable-diesel systems ranging from 150 kW to 200 kW. This component was targeted to benefit 200,000 herders through improved electricity services offered at *soum* centers by the end of 2011.

Institutional Capacity Building

This component was crafted to strengthen the capacity of the NREC to carry out the project. This included preparing business plans and work programs, project monitoring and evaluation, and reporting. It also encompassed the provision of technical and management training for NREC staff, with the idea that such capacity-building could assist Mongolia in developing a more robust and effective regulatory framework at the national level for promoting renewable energy.

Part of the capacity building also included extensive consultation with donors, banks, and solar and wind equipment suppliers. Funds were provided to host a 2005 workshop before REAP officially began so that ideas could be brainstormed, and a Mongolian delegation visited China to learn about renewable energy deployment to rural areas there. Funds were disbursed to host an international conference in Ulaanbaatar convening bankers, donors, and energy experts to discuss rural credit institutions, renewable energy, and private manufacturing.

Benefits

The REAP program has so far yielded four distinct benefits: expanded access to energy supply, improved quality of SHS, improved affordability of energy services, and reduced greenhouse gas emissions.

By far the most direct and substantial benefit from REAP, as stated in World Bank objectives, is expanded access to electricity for communities that had previously "fallen through the cracks." Herders have seasonal products to sell and thus experience fluctuations in income, offering cashmere in the spring, meat and livestock in the winter, and cheese, milk, yoghurt, and curd in the summer. Since banks would not accept livestock or livestock products as legitimate collateral, as cows and horses could die, freeze, or spoil in harsh winters, and due to the difficulty of tracking herders down, they were often rejected for loans to pay for the capital costs of SHS and WTS. REAP changed this and partnered with banks so that herders could buy renewable energy technologies in interest-free installments.

As of early 2010, when our field research concluded, REAP had already distributed 41,800 SHSs and a few hundred WTSs. The typical SHS costs $330, including transportation, and with a project subsidy of $160, the herder pays only $170. (SHS have proven much more popular than WTS due to their ease of assembly and disassembly, important for nomads on the move, as well as the greater reliability of solar resources compared to the site specific nature of wind resources). The officials we spoke with at the World Bank and Mongolian Energy Authority expect the 50,000 target to be reached ahead of schedule and if the government is able to procure an additional 20,000 SHS as planned, to exceed that target by some 10,000 SHS by early 2012. REAP has also facilitated the rehabilitation of 15 out of the 30 mini-grids in *soum* centers and installed 11 out of 20 renewable-diesel hybrid systems and seen five other *soums* connected to the main power grid.

Herders primarily used the electricity from these SHS, micro-grid, and hybrid systems for lighting, satellite television, and radio, but also occasionally for cooking and refrigeration when they purchased larger units and/or sufficient battery storage. Figure 6.2 shows a girl near Nalikh watching television powered by her SHS. The families we spoke with talked about the "huge difference" a SHS could make for them, enabling them to keep meat and milk cold during the summer and improve their livelihood by making reading possible at night and utilizing radios and television programs to be informed of current events. Some also used the electricity to charge mobile phones that alongside television offered real-time trading prices for cashmere, dairy, and meat products, enabling herders to more profitably participate in local markets.

A second benefit was improved service quality of renewable energy systems. While sales centers existed before REAP's introduction, they were primarily located in urban centers and therefore unsuitable for providing accessible after-sales support for herders in remote areas. REAP created dozens of new centers to help herders maintain their systems, provide advice on battery charging, distribute spare parts, and honor warranties. These new centers served as focal points to enforce newly established equipment standards and have conducted wind and solar resource assessments for *soum* centers wishing to deploy hybrid electricity systems. They typically sell bundled items such as SHS-televisions, SHS-appliances (such as the electric milk separator used to make butter), and SHS-DVD players that operate using DC electricity and do not require an inverter.

The project sought to extend the reach of existing centers, as well as foster the creation of new ones. There are now more than 30 private Mongolian companies distributing PV equipment with a handful of larger firms who have previously worked with the government. REAP staff anticipate the total to reach 70 before the close of the project.

Figure 6.2   An electrified *ger* running a color television, near Nalaikh, about 60 km east of Ulaanbaatar, Mongolia

Over the course of REAP, unit costs for SHS have also declined and quality has improved, with systems lasting two to three years longer than before and costs dropping an average of 5 percent each year (according to one interview respondent). Higher quality systems also helped mitigate the sales of black market items that would falsely claim higher wattage capacities. Since Mongolia lacks the appropriate facilities, SHS were sent to China for testing as one measure to introduce product quality standards, compliance with those standards, and warranty requirements. REAP also sponsored a "Renewable Energy Road Show" where a team of project personnel traveled around different *soums* demonstrating the benefits of SHS, offering information about prices, and showing herders the different applications of electric appliances (including hair driers and electric weavers alongside computers, televisions, and milk separators).

A third benefit was greater affordability of energy services. Before REAP energy services were provided mostly by old, polluting, inefficient diesel generators.

These generators not only relied on expensive diesel fuel, but also tended to breakdown frequently and perform poorly under the extremely low temperatures of the Mongolian winter.[19] Villagers and herders that we spoke with commented that the price of diesel electricity was four to five times the price of grid electricity and about twice as expensive as the electricity from an SHS (though such prices would conceivably include the distortionary effects of any related subsidies for either source). As one sign of the economic viability of the new SHS, a majority of herders paid the total SHS costs upfront. Others utilized existing banks with branch locations in rural areas to take out a one-year, no-interest loan.

A fourth benefit is avoided greenhouse gas emissions. The dozens of towns and villages not connected to the grid rely on diesel generators that emit prodigious amounts of greenhouse gases, and *gers* use large amounts of coal for heating and cooking. SHS and WTS frequently displace the need for these diesel generators completely and partially displace the need for coal- and dung-based cooking where electric cookers and kettles are used, though these latter energy services typically require capacities above 100 Wp for SHS and 1 kW for wind turbines. One assessment calculated that the REAP project will avoid the emission of 184,000 metric tons of CO_2 from 2006 to 2020 in comparison to the project not taking place.[20]

Challenges

Notwithstanding these benefits, REAP faces some remaining challenges. These include the upfront cost of renewable energy systems, dependence on imported materials and technology from China, remaining lack of institutional capacity, lingering obstacles related to improved consumer awareness, and a political commitment to fossil-fueled and centralized energy systems.

Even with the financing provided by REAP, herders still have to provide 50 percent of the system costs themselves. With many making only $300 to $400 a year, this can amount to half of their annual income. Even before the project commenced, surveys with herders found that most would not be able to purchase even a basic 20 Wp system with financing and that most herders expected more than just lighting from an SHS, meaning they would want far more than a small 20 Wp device.[21] The winter of 2009 was also unusually cold. Nomadic families lost more than 4.5 million livestock and therefore have fewer assets on which to draw from to pay for SHS and WTS.[22]

---

19   GEF 2006b.
20   GEF 2006b.
21   GEF 2006a.
22   *Economist* 2010. Bitter Toll: Mongolia's Zud. April 3, 44.

It also remains unclear whether REAP truly benefitted the private sector. A healthy number of solar and wind distributors existed before the project began[23], but many were not involved in the program. We know of at least two that went out of business because they could not outperform REAP-subsidized technologies. As one private sector participant noted, "it is difficult to compete with a government subsidized program." Moreover, none of the 17 manufacturers of SHS in Mongolia actually build them in the country; Mongolian retailers instead import panels from Chinese manufacturers such as Beijing Hub and Fivestar Solar. The REAP decided to sidestep retailers and do bulk purchases of imports to get better prices, but may have hindered the long term relationship that individual retailers enjoyed with suppliers in China. By accelerating subsidized sales of SHS in a short period of time, the program may have preempted the development of a steady SHS market, leaving retailers to worry about the long-term sustainability of SHS business. (Sadly, there is no way of rigorously quantifying what happened before and after the introduction of the program due to lack of reliable data).

Institutional capacity is still lacking within and between government planners, development agencies, *soum* centers, and retailers. Some interviewees stated that the "institutional and regulatory vacuum" that was supposed to be filled according to World Bank's pre-project documentation still remains, with no single government agency committed to advancing renewable energy. Current Mongolian energy planning processes also suffer from inconsistent policies and poor coordination between different ministries and a lack of transparency. We experienced some of this personally with attempts to speak with representatives from MOFE who repeatedly declined (before one eventually agreed to speak to us via telephone). There were also gaps in the evaluation and monitoring of the program, with no studies done yet looking at the value added from REAP to the livelihoods of herders.

As one example of inconsistency, interview respondents noted that planners in Ulaanbaatar expect *soum* utilities to provide electricity to public institutions and families at a profit, but also constrain their ability to collect bills and install meters. One company challenged components of the REAP project on the grounds that they were incompatible with Mongolian law about barriers to entry and exit from the renewable energy market, its case going all the way to the Supreme Court. The Renewable Energy Law of 2008 establishes a feed-in tariff with preferential rates to fund renewable energy systems (giving 8 to 12 ¢/kWh for qualified systems when the going market rate for electricity is 5 ¢/kWh), but so far has not implemented it and awarded no funds for the disbursement of the tariff.

Additionally, awareness efforts still have much to accomplish. Many herders still lack information and understanding about what SHS can do, and also about the REAP program in particular. As one respondent explained, "if one goes to very remote areas in Mongolia and speak with herders that have purchased a SHS,

---

23 Lin, L. 1999. Renewable Energy Utilization in Inner Mongolia in China. *Renewable Energy* 16, 1129–1132.

Figure 6.3   **A typical fuelwood cookstove in Mongolia**

they will say they bought it because a neighbor did, and not because they were convinced about the merits of solar energy from the advertisements and brochures produced by the government, or because they thought system could really benefit them." The point here is not that word of mouth advertising is somehow bad, but that some users of SHS cannot articulate the reasons behind their decision to purchase them. Some herders have voiced unrealistic expectations about SHS' capacity to power an entire *ger* replete with lavish appliances, a television, refrigerator, and electric stove. As another respondent stated, "there are certainly some herders that think solar panels can do everything, including cooking," when in reality most cooking is done through the fuelwood cookstoves shown in Figure 6.3, rather than electric appliances.

Lastly, the government appears divided about its overall national energy policy and the role of renewable energy. Some officials place a low priority on rural electrification efforts and off-grid systems, and instead see fossil fuels, nuclear power, and centralized electricity supply as a better investment. Others desiring to promote renewable energy dedicate their efforts to larger-scale commercial wind farms, hydroelectric dams, geothermal projects, and concentrated solar facilities. While we agree planners in Mongolia ought to be doing both urban and rural electrification, all too often stakeholders argue only in favor of one or the other.

For example, MOFE has placed considerable emphasis on further expanding the production of coal in Mongolia, both for domestic use and exports. Table 6.4 shows that the production of coal is expected to more than double from 27 million tons today to 63 million tons by 2015, placing an abundance of cheap coal on the local market for use by power plants and herders, who will legitimately need some solid fuel for heating and cooking. The government has also announced plans to build a new 480 MW coal-fired facility near the Shivee Ovoo mine to provide electricity throughout the area as well as a 150 MW plant near Uhaahudag and an additional 450 MW facility at Oyu Tolgoi. With annual economic growth rates forecast to surpass 10 percent from 2011 to 2013, the government also expects that they will need to build more centralized plants around Ulaanbaatar and extend the grid to accommodate new commercial and industrial users. As one respondent remarked, "Coal will continue to be the main source of energy in Mongolia's near future."

Others have touted nuclear energy as the way to go, with the Parliament establishing a state owned company called "Mon-Atom" to conduct geological surveys of potential plant sites and to explore recoverable domestic uranium resources. These efforts would all attempt to extend the centralized electricity grid to rural areas rather than disperse small-scale, decentralized sources of energy close to *soums* and *gers*.

Table 6.4   Major existing and planned coal mines in Mongolia, 2009

Mine	Life	Production (million tons/year)	Employees	Start date
Nariin Sukhait	40	12	150	2003
Ovoot Tolgoi	50	5	400	2008
Uhaahudag	40	10	1,000	2009
Tavan Tolgoi	200+	15	1,500	2012
Baruun Naran	20	6	500	2012
Tsagaan Tolgoi	20	2	150	2015
Sumber	50	5	400	2015
Shivee Ovoo	200+	8	600	2015

Furthermore, those within the government committed to renewable energy have harnessed their efforts on larger rather than smaller systems. Hydroelectricity, both large and small, is seen as an important way to provide electrification in the North and Southeast, and investors from China, Germany, and Japan, along with the International Finance Corporation, have already put their money into a 50 MW wind farm near Salkhit (consisting of sixty 850 kW turbines), another 50 MW wind farm near Sainshand, and a 250 MW wind farm in the south near Khanbodg. If the government is to meet its renewable portfolio standard target of 20 to 25 percent by 2020, one respondent suggested that "we will need to deploy renewables in large, multi-MW chunks, not eat away at the margins with systems measured only in watts." Another respondent suggested that "without special subsidies and interventions aimed at small-scale, rural users, there's no chance SHS and WTSs can compete with these larger renewable energy systems."

Similarly, those within the Ministry of Environment display a predilection for projects that can generate a large volume of Certified Emissions Reductions under the Clean Development Mechanism (CDM) of the Kyoto Protocol, such as building commercial wind farms, deploying geothermal power stations, upgrading grid infrastructure, designing combined heat and power systems, implementing energy efficiency efforts, and refurbishing hydroelectric dams, not distributing SHS and WTS. And planners in Ulaanbaatar are committed to tackling the growing problem of air pollution and air quality and poverty within urban areas, a problem so acute because people "cut down trees and even burnt their tires and furniture to keep warm last year." These practices resulted in levels of ambient pollution that far exceed European or North American standards.

In essence, a government planning to build hundreds of MWs of coal-fired capacity to deliver electricity to urban commercial and industrial users is not likely to place a high priority on further expansion of SHS and WTS, given the sentiments of the officials we interviewed and spoke with. Those within the government committed to renewables orient themselves towards commercial-scale projects that can help the country meet its renewables targets and generate CDM credits. Neither is enthralled with the small customer base of off-grid nomadic herders and isolated *soum* centers.

Conclusion

While constrained by high upfront system costs, gaps in institutional capacity and public awareness, and a political commitment to conventional forms of electricity supply, REAP did a remarkable job distributing 41,800 SHS and hundreds of WTS to rural users in Mongolia. It expanded electricity access for herders and rural households so that they could receive lighting, warmth, comfort, refrigeration, and entertainment either for the first time or without relying on expensive and polluting diesel generators. REAP improved the quality of solar and wind technology by extending the reach of service centers and creating minimum standards. It will displace 184,000 tonnes of carbon dioxide emissions over the next 15 years from SHS and WTS already installed. If the program continues on its current trajectory, it will reach 60,000 SHS distributed by the end of 2012, ten thousand more than its original target.

For planners in Mongolia, REAP serves as an enduring reminder that centralized and capital intensive energy systems may not always be the cheapest or most environmentally benign way of providing remote communities with energy services. Indeed, it remains unlikely that existing plans to expand the grid into rural areas will ever be able to truly serve the needs of nomadic herders, who will continue to reside with no fixed location. This suggests that grid electrification efforts should continually be complemented with targeted policies and programs aimed at assisting these herders. The Mongolian market for SHS and WTS also seems to be in its nascent stages, currently dependent on imported Chinese systems but also with the potential to become a robust and self-sustaining part of the local economy with the right amount of support. If Mongolia takes its commitments to provide all rural households with energy services seriously, it should consider extending and expanding REAP.

REAP suggests that rural energy programs work best when they ensure the participation of the private sector and also solicit feedback from consumers and end users. It took an integrated approach to coupling rural electricity services with development and focused not only on deploying SHS and wind units to nomadic herders, but also on rehabilitating electricity grids and shifting from diesel-produced electricity to hybrid diesel-renewable and fully renewable energy micro-grids. The program undertook efforts to improve technology through certification, standardization, and after-sales service alongside efforts to improve regulatory frameworks, build institutional capacity, train workers, and demonstrate solar and wind applications for herders.

It also, finally, implies that efforts at rural electrification, especially for small and targeted communities such as herders, will not always be profitable. As one respondent told us, "the program would not work without subsidies and policy intervention, coal and conventional energy is just too cheap and available, and without outside subsidies the program would have collapsed." The lesson here implies that for successful renewable energy development to occur, achieving volumes of sales may need to be placed as a greater priority than acquiring profits in the short term. When these separate pieces—participation, private sector engagement, integration, technological learning and improvement, an orientation to public service—come together for a single project, they do indeed seem to overcome the usual political, economic, and technical problems that sometimes plague other programs.

Chapter 7

The Rural Energy Development Project in Nepal

Introduction

For those that have only heard of Nepal in fables and travelogues, it is a landlocked country in South Asia bordered by China and Tibet to the north and India on its other borders. The Himalayan mountain range crisscrosses the northern and western parts of the country and it is home to eight of the world's ten highest mountains, effectively isolating Nepal and earning it the nicknames "the roof of the world" and "the landlocked island."[1]

Within a few hours of touching down in the Kathmandu Valley, the extent of Nepal's electricity crisis becomes painstakingly apparent. Upon entering the capital city of Kathmandu, the air pollution, mostly the result of combusting fossil fuels, became so pungent it made our eyes sting, we could taste it in our mouth, and pedestrians could rarely be seen crossing the street without wearing protective face masks. Upon arriving at our hotel, the electricity promptly went out. When the clerk was asked how long the current outage would last, he shrugged and handed one of us a candle. When walking around the bazaars later that night, we could see that more shops relied on candles to light their wares than electric bulbs, one of them depicted in Figure 7.1.

Indeed, though Nepal has a population of about 30 million people, one-third of it is officially "poor," and on the United Nation's Human Development Index it ranked 144 out of 179 countries. Nepal has strikingly low levels of access and electricity consumption, with an average of only 86 kWh consumed per person annually compared to the global average of more than 3,000 kWh. Only 40 percent of the rural population has access to electricity, and hydroelectricity and coal each meet only one percent of national energy needs.[2] For many rural communities, access from the nearest road is only possible by foot or mule and can take several days. On average, Nepalese women spend 41 hours per month collecting fuelwood and some 12 hours per day performing daily cooking and household chores.[3]

---

1 Robin Shields 2009. The Landlocked Island: Information Access and Communications Policy in Nepal. *Telecommunications Policy* 33, 207–214.

2 Clemens, E. et al. 2010. *Capacity Development for Scaling Up Decentralized Energy Access Programs*. Kathmandu: UNDP, AEPC, and Practical Action Publishing.

3 Bastakoti, B.P. 2006. The Electricity-livelihood Nexus: Some Highlights from the Andhikhola Hydroelectric and Rural Electrification Centre (AHREC). *Energy for Sustainable Development* 10(3), 26–35.

Figure 7.1　A butcher selling fish by candlelight in Kathmandu, Nepal

Such poverty, lack of access, and gender disempowerment make it all the more notable that a recent national project designed to bring rural communities energy services has surpassed many of its targets. This chapter explores the contours of the Rural Energy Development Program (REDP), a scheme funded by a consortium of multilateral donors, including the World Bank, the United Nations Development Program (UNDP), and the government of Nepal, to develop small- and medium-scale microhydro energy units. The REDP in essence installed small grid-connected hydro where relevant and microhydro for villages off of the grid, and this chapter explores how it accomplished its goals.

The primary value of our exploration extends beyond Nepal. The World Bank has noted that for off-grid communities around the world, hydroelectric systems (where this resource is available) can offer the *cheapest* generation costs compared to every other commercially available technology of the same size, including solar home systems (SHS) and diesel and gasoline generators.[4] Yet for a variety of reasons, such systems continue to achieve only a small fraction of their potential in most developing countries. Our chapter offers an in-depth exploration of what aspects of the REDP are replicable for other countries, exposing how many of the barriers facing grid-connected and off-grid hydroelectricity can be overcome.

4　World Bank 2007. Technical and Economic Assessment of Off-Grid, Mini-Grid and Grid Electrification Technologies. *ESMAP Technical Paper 121/07.*

Description and Background

Life in Nepal is mostly agrarian, with three-quarters of all Nepali people working in the agricultural sector. Energy consumption differs from that in most other countries in two obvious ways: modern energy carriers such as petroleum, coal, and hydroelectricity meet about 10 percent of national energy needs, and the residential sector accounts for 90 percent of overall consumption. Estimates suggest that due to its rugged and mountainous terrain, Nepal has 83,000 megawatts (MW) of exploitable hydropower resources (and 43,000 MW of "economic potential" or 180,000 GWh/yr), flowing from its 6,000 fast flowing rivers and streams totaling more than 45,000 kilometers in length.[5] Yet, so far less than one-tenth of a percent of this potential has actually been tapped.

Most of Nepal's electricity generating capacity is concentrated near and serves Kathmandu, with about 53 megawatts (MW) of thermal electric stations operating on diesel near the city center and a collection of hydroelectric dams spread across the northern mountains. Pumped storage supplements another 92 MW from the Kulekhani Dam, but the ability to reliably operate all types of hydroelectric power stations varies greatly with the seasons, with about 480 MW dependable in the wet season but only 190 MW available in the dry season, when waters in the rivers greatly recede. Although Nepal's total installed capacity is therefore about 700 MW, when independent power producers are added to the mix, demand for electricity surpassed 970 MW in 2009, responsible for the daily 16 hours of load shedding described in the introduction.

The story of small-scale hydroelectricity in Nepal started in the late 1960s, when international development donors ran pilot projects with local manufacturers to test the use of microhydro units for mechanical agricultural processes (mainly used for grinding, husking, and oil expelling). The 1970s witnessed a slew of workshops hosted by Swiss and Norwegian experts on how to develop and test equipment and in 1984, the government began nationwide efforts to promote medium, small, and microhydro systems in both grid connected and off-grid configurations. The Seventh Five Year Plan (1985 to 1990) argued that the provision of basic needs for the whole population was of key importance, and in 1989 the government legislated that all rural electrification projects were eligible for a 50 percent subsidy, rising to 75 percent for remote installations. Of those installed at the beginning of the early 1990s, more than 90 percent were primarily for mechanical power for processing, with only a quarter having any application

---

5   See Nepal Ministry of Water Resources and Ministry of Population and Environment 1997a. *Nepal Power Development Project Sectoral Environmental Assessment Volume 1*. Kathmandu: Government of Nepal; and Dhakal, S. and Anil K.R. 2010. Potential and Bottlenecks of the Carbon Market: The Case of a Developing Country, Nepal. *Energy Policy* 38, 3781–3789.

for electricity. Ninety percent also received financing from a single government actor, the Agricultural Development Bank of Nepal.[6]

During the 1990s, however, the microhydro power sector in Nepal entered what one respondent called "a downward spiral." This main reason for this was the difficulty in operating and maintaining microhydro systems, leading to widespread frustration and loss of revenue in most projects. A majority of these projects failed to cover even their operating costs. Nepal Electricity Authority (NEA), the national utility, was obliged to meet shortfalls by subsidizing projects, draining government revenue. In contrast with hydroelectric sectors in other economies, there was little standardization between systems because those in Nepal were almost entirely built with the support of development donors. Different tenders, technologies, and contractors made few projects alike, making servicing, maintenance, and repair difficult. Many schemes were also designed and planned without information about hydrology, geology, or community economies to quickly take advantage of subsidies before they expired. This led to "poor siting, unsuitable designs, incorrect projection of demand and interference with water requirements for irrigation,"[7] resulting in a lack of readily adaptable technology of consistent technical standards, and dependence on external sources of parts and expertise.

NEA, which did have support staff, had little incentive to manage projects successfully, as its revenues were not kept within the organization itself but passed to the Ministry of Finance. Even small purchases of spare parts required a lengthy application and approval process. The confluence of these factors meant that in the 1990s, "investments in microhydro plants [were] stagnating or falling and existing plants experience[d] operational and institutional problems."[8]

Things worsened during the same period with financial troubles at NEA, and constrained power exchanges and trading with India, leading to poor quality of electricity supply with high system losses and overall high costs of power outside the range of most of the population, especially the poor.[9] More than three-quarters of all microhydro plants had overdue loans, about one-third were not operating due to floods and landslides, and two-fifths of service centers set up to do hydropower maintenance had closed. Although the Alternative Energy Promotion Center (AEPC) took over the administration of renewable energy, including microhydro,

---

6   See Cromwell, G. 1992. What Makes Technology Transfer? Small-Scale Hydropower in Nepal's Public and Private Sectors. *World Development* 20(7), 979–989; Pandey, B.R. and Cromwell, G. 1989. Cost-Effective Monitoring of Technology Transfer: Case Studies from Nepal. *Appropriate Technology Journal* 16(4), 101–111.

7   Cromwell 1992, 985.

8   Mostert, W. 1998. Scaling Up Micro-Hydro: Lessons from Nepal, presentation to *the Village Power 1998 Scaling Up Electricity Access for Sustainable Rural Development Conference*, Washington, DC, October 6–8.

9   World Bank 2003c. Project Appraisal Document on a Proposed IDA Credit to Kingdom of Nepal for a Nepal Power Development Project. *Report Number 23631-NP.* Washington, DC: World Bank Energy and Infrastructure Sector, South Asia Regional Office.

in 1996, by 2003 only 200,000 rural households had been electrified (meaning the national rural electrification rate was just 5.7 percent) with millions in rural areas unlikely to receive electricity in the foreseeable future.

To address these challenges, the World Bank and United Nations Development Program, in conjunction with local banks and the government of Nepal, scaled up the REDP in 2003 with the objective of improving access to rural electricity services through microhydro units. Planners deemed the "traditional" strategy of promoting large-scale hydroelectricity "infeasible" because of the difficulty with raising large amounts of capital for the upfront costs of big dams, and the prohibitive nuisance of reaching remote communities by extending the national grid. Though it overlapped with another ongoing effort called the Nepal Power Development Project, the REDP sought to expand service coverage in rural areas so that 30,000 new households were provided with electricity services and 125 to 150 new microhydro systems would be constructed by 2008.[10]

Essentially, the REDP was designed to accelerate community-level projects with a total capacity of 2.5 to 3.0 MW, serving about 30,000 new customers and 10 new districts. It supported hydroelectric facilities ranging from 10 kW to 100 kW, with an average plant size of 25 to 30 kW. Program implementation was decentralized to local governments, with District Development Communities (DCCs) and Village Development Communities (VDCs) required to form Microhydro Functional Groups in each community. The AEPC, an autonomous body established in 1996 under the Ministry of Science and Technology, was tasked with assuming overall management. The primary benefit was to provide customers currently dependent on kerosene and other fuels for lighting with reliable electricity. A secondary benefit came from the promotion of end-use activities such as cereal milling, rice husking, and mustard seed processing as well as the replacement of manual implements for carpentry by electrical machines and tools—though to prevent deforestation, project financing could not be used for commercial sawmills. The REDP required that communities wishing to build microhydro facilities donate land for the construction of canals, penstocks, power houses, and distribution lines voluntarily. Furthermore, villagers were required to contribute land and labor for civil works related to microhydro units.

Though microhydro schemes were exempt from the Environmental Protection Act 2053 (1996) and Environmental Protection Rules 2054 (1997), and thus not required to conduct environmental impact assessments, specific evaluations still had to be undertaken in accordance with the guidelines established by the REDP. These standards strongly emphasized the mitigation of environmental impacts,

---

10   See United Nations Development Programme 2008. *Impacts and Its Contribution in Achieving MDGs: Assessment of Rural Energy Development Program*. Kathmandu: UNDP's Rural Energy Development Program; World Bank 2003a. *Integrated Safeguards Data Sheet*. Nepal Power Development Project. Washington, DC: World Bank, Report Number AC76; and World Bank 2003b. *Updated Project Information Document (PID)*. Nepal Power Development Project. Washington, DC: World Bank, Report AB30.

ensuring that residual flow in the dewatered section of rivers was never less than 10 percent of the dry season flow. They also mandated that lined channels and pipes flush flows back to rivers, that canals were lined to prevent leakage, and that slopes were carefully excavated to minimize landslides and subsidence.[11]

Tariffs for microhydro units were set by each Microhydro Functional Group, and were based on loan repayments, operation and maintenance costs, depreciation, and provision of a reserve fund for maintenance. Only schemes expected to yield average economic return of 10.9 percent for the program as a whole, or individual returns ranging from 10 percent to more than 12 percent, were supported.

One unique element of the REDP was its emphasis on maintenance. The program provided extensive training in operations and maintenance for local operators and managers from each community doing a microhydro project assigned to each system, so that they would understand technical aspects of system operation, bill collection, disconnecting for non-payment, record keeping, and accounting. Turbines and generators were to be manufactured in Nepal, and maintenance support facilities and service centers within districts were to be established and strengthened to provide repayable financial support.

Another unique element was that part of REDP project funds was given to promote women's empowerment, skills enhancement, better management of technology, and income generation in a "Community Mobilization Fund" (CMF). This fund focused on coupling hydropower with income generation schemes, and it offered $400,000 in total for the promotion of non-lighting uses of electricity such as agro-processing, poultry farming, carpentry workshops, bakeries, ice making, lift irrigation, and water supply. To support these activities, local CMFs were established to cover the financing of local projects. The CMF also gave grants for power connections from microhydro schemes to schools, health posts, clinics and hospitals, and it promoted afforestation to offset any trees felled for the construction of distribution poles. CMF funds were lastly utilized for community training and to educate the operators of microhydro plants and other end-use machinery.

Due to a variety of factors, including a protracted civil war, the REDP was restructured in 2008 and 2009. In essence, the REDP was redesigned so it no longer had to rely on the government to meet its remaining targets; it was "simplified" so that it could be done solely by other actors. Also, with the injection of more funding, REDP targets were extended upward to include the building of an additional 1.5 MW of hydroelectricity capacity to serve 15,000 more households and activities in 32 more districts. The closing date of the project was also extended to December 31, 2012.

---

11   United Nations Development Programme 2005. *Environmental Assessment Guidelines for Community Owned and Managed Micro Hydro Schemes*. Kathmandu: UNDP's Renewable Energy Development Program.

Benefits

Under the REDP, microhydro system coverage grew from only a few thousand homes in 25 districts in 2003, to 40 out of 51 targeted districts and 40,000 households in 2007, and more than 50,000 homes in all targeted districts as of November 2010. The number of microhydro projects also jumped from 29 in 2003 to 280 in mid-2010, as Figure 7.2 shows. As of December 2007, when the World Bank did a formal evaluation, a total of more than 90 projects with 1.5 MW of capacity had been completed, providing access to 16,914 households, meaning original project deadlines had been completed almost 18 months ahead of schedule. By the end of 2012, project managers told us they expect to be operating in all 75 districts with 6 MW of capacity installed reaching more than one million people.

As of November 2010, when we spoke with project managers at the AEPC and UNDP, plans were underway to expand service coverage to 30,000 new households and construct 125 to 150 new microhydro systems by 2012, adding a potential 5 MW more. Efforts were also underway to integrate microhydro schemes with the Clean Development Mechanism of the Kyoto Protocol so communities could earn Certified Emissions Reductions credits, and the government was discussing a three-year interim plan to place a microhydro unit in *every* northern village (microhydro units are feasible only in the hilly and mountainous north, not the flat plains of the south). A pilot project in Baglung district created a mini-grid consisting of several microhydro plants stitched together. Research was also being conducted to double the load factors for existing microhydro units and to focus on connecting some of them to the grid so they could make additional revenue exporting electricity.

Respondents noted how the hundreds of microhydro units installed under the scheme provided not only lighting but also "mechanical energy for milling, husking, grinding, carpentry, spinning, and pump irrigation that have paid off in higher local incomes." Figure 7.3 shows one such microhydro irrigation system near Kavre. As one participant explained:

> Microhydro units are much easier to operate, cleaner, safer, and cheaper than the diesel units they often replace. You can train local people without any technical background to operate and maintain a microhydro system. It's not a hard thing to learn, it can be easily handled entirely by local communities. They can maintain it themselves and collect tariffs themselves. In Lukla, where we installed a 100 kW system, people used to sleep at 6pm or 7pm, after dark, now they can work after 10pm, do more work, study longer hours, in short they can live a more productive life.

Respondents noted that microhydro units are also able to provide electricity in remote mountain areas unsuited for biogas (because fermentation takes more time at higher altitudes) and solar home systems (because of consistent fog and cloud cover).

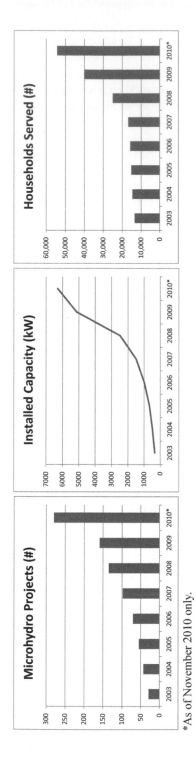

*As of November 2010 only.

Figure 7.2   Microhydro projects, installed capacity, and households served under the REDP in Nepal

Figure 7.3   A 15 kW microhydro system in Kavre, Nepal, provides irrigation and productive energy in addition to electricity supply

One of the benefits of the microhydro units is that insurgents did not target them during the protracted Maoist civil war, in part due to the extensive involvement of grassroots village communities. As one respondent elaborated:

> Not a single microhydro unit under the entire REDP was ever bombed or extorted. Commercial systems were attacked, destroyed, sabotaged, blockaded, and protested. The Maoists realized they would quickly alienate the people they were trying to protect if they were to start targeting community-based microhydro units. So they left them alone.

Another noted that "the whole country came to a halt with the civil war, but we didn't have a single problem with the REDP." As such, while the World Bank expected five to 10 percent of all microhydro units under the REDP to "fail," due in part to lack of maintenance but also the uncertain political environment[12], respondents told us that 99 percent of all plants are still working technically (with a failure rate of "less than 1 percent" or "less than 5 percent," depending on who we asked) and "100 percent are financially sound, that is they have paid off their loans or have not missed a single loan payment."

Respondents also affirmed that the CMF funded "a variety of income generating activities" and "trained and empowered thousands of women and villagers." One noted that:

---

12   World Bank 2003a; 2003c.

Table 7.1 **Energy use for microhydro and non-microhydro rural households in Nepal (%)**

Energy source	Cooking	Lighting	Heating	Electric appliances	Home business
Microhydro users					
Fuelwood	100	0	0	0	0
Dung	1	0	0	0	0
Straw	4.7	0	0	0	0
Kerosene	1	72.6	0	0	0
LPG	0.5	0	0	0	0
Charcoal	0	0	0	0	0.3
Coal	0	0	0	0	0.6
Solar home systems	0	0	0	0	0
Biogas	6.2	0	0	0	0
Microhydro	0.5	100	0	9.5	0
Candles	0	16.2	0	0	0
Dry cell batteries	0	58.8	0	4.1	0
Non-microhydro Users					
Fuelwood	100	0	21.9	0	0
Dung	0.1	0	0	0	0
Straw	0.7	0	0	0	0
Kerosene	3.9	84.8	0	0	0
LPG	0.2	0.04	0	0	0
Charcoal	0	0	0	0	0.3
Coal	0	0	0	0	0.2
Solar home systems	0	5.3	0	0	0
Biogas	5.9	1.3	0	0	0
Microhydro	0	0	0	0	0
Candles	0	11.1	0	0	0
Dry cell batteries	0	78	0	17.2	0

The CMF has successfully created a revolving national fund that provides much needed resources for community development. Since it mostly gives low interest loans rather than grants, it is replenished each year as projects pay back into the scheme. It has so far helped hundreds of communities couple microhydro electricity production with income generation and empowerment.

As another respondent noted "It may be small hydro, but it is big impact." Various studies have attempted to quantify the benefits of the REDP program, and have come to consistently positive results. World Bank experts told us they calculated an average cost of the REDP program per household at $1.40 per month but net benefits of $8 per month.[13] Respondents at the AEPC were less optimistic but said a "worst case scenario" was $1.60 back for every $1 invested into the program.

Three more formal evaluations have reached similar conclusions. One study surveyed 2,500 households and 70 commercial enterprises in Nepal, including a mix of those connected to microhydro schemes and those without them, as well as a representative sample of control groups from all five development regions (East, Central, West, Midwest, Farwest), to assess the benefits of microhydro systems installed under the REDP offered to communities.[14] It documented that microhydro users consume far less kerosene than their non-microhydro counterparts; that communities with microhydro units have comparative increases in income by 11 percent; and that women and children from these communities suffer less from respiratory problems and incidences of disease. As Table 7.1 shows, non-microhydro households also relied more on expensive dry cell batteries and fuelwood for lighting and cooking. The study also calculated that microhydro households emit about 3.6 kilograms of carbon dioxide less than their counterparts, meaning microhydro units displace nearly 10 million kilograms of carbon dioxide each year.

The UNDP conducted a comparable evaluation of 20 microhydro project sites involving 2,543 questionnaires completed by 1,503 households in 10 districts.[15] Their study found that microhydro communities saw significant growth in household income, with an average increase of 52 percent from 1996 to 2005, much higher than the national average. They noted that microhydro communities had higher literacy rates, with 90 percent of households saying

---

13 Based on the assumption that (1) An average household has a connected load of 100 Watts (W); (2) consumption at that household is about 6 kWh per month; (3) kerosene consumption is 3.5 liters per month per household; and (4) average equipment lifetime is 15 years before major refurbishments are needed.

14 Banerjee, S., Singh, A., and Samad, H. 2010. *Power and People: The Benefits of Renewable Energy in Nepal*. Washington, DC: South Asia Energy Unit, World Bank.

15 United Nations Development Programme 2008. *Impacts and Its Contribution in Achieving MDGs: Assessment of Rural Energy Development Program*. Kathmandu: UNDP's Rural Energy Development Program.

lighting and microhydro services contributed towards positive educational change. Microhydro communities had more equal gender roles, with households reporting a sensitization of communities to the value of women entrepreneurs, with changes in social perceptions, income generation for women, improved sanitation, and health and time savings. The study also documented less direct expenditures on energy services, with savings of 29,000 liters of kerosene per year from the 1,503 surveyed households, as well as a reduction in batteries needed per month. Interestingly, the study noted less firewood collected in microhydro households as electricity displaced the need for wood-based lighting, with monthly demand for firewood dropping from 10 bharis in 1996 to 7 bharis in 2005.

Lastly, a final study looked at the microhydro sector in Nepal from 1996 to 2006 and noted that the REDP had driven significant improvements in technology, with a total program cost per installed kW falling 73 percent from 1996 to 2006.[16] The study also noted that capital costs dropped from $17,300 per installed kW in 1996 to $4,600 per installed kW in 2006 (and even lower since then), hardware costs declined 33 percent, and capacity development costs declined 84 percent, as Figure 7.4 shows. The study demonstrated that the average expenditure for the entire program amounted to about $110 per household on average in 1996, but had dropped to $70 per beneficiary going forward in 2010. The REDP program had also become more "sustainable" and less reliant on government financing. When it started, more than 90 percent of project funds had to come from public sources, but an average project in 2010 was funded only with nine percent of public support

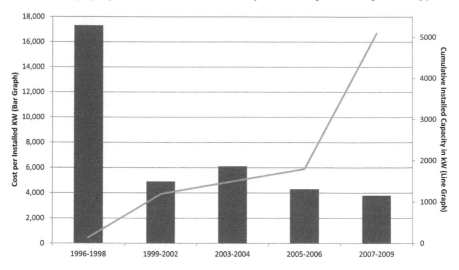

Figure 7.4   Installation cost and cumulative capacity of REDP microhydro units in Nepal, 1996–2009

---

16   Clemens et al. 2010.

and community expenditures above 75 percent (the rest coming from banks and commercial lenders). The specific REDP program was also complemented with being responsible for 40 percent of all national microhydro systems installed.

Challenges

The REDP faced, and still faces, some technical, economic, political, and social obstacles.

Technical challenges relate to design and project siting, maintenance, and manufacturing. In terms of design, the prevalence of early vertical access water wheels meant that Nepali microhydro units often relied on cross flow turbines, not really designed for electricity generation but adapted from milling. These turbines have natural limits to their efficiency, operating at only 60 to 70 percent the efficiency of other designs. Complicating matters, site specifications like head, flow canal length and accessibility to roads differ from site to site, making construction difficult, and also creating a large variation in project costs (ranging from a low of $1,200 per installed kW to a high of $18,000 per installed kW).[17]

Respondents also noted that "during the rainy season, engineers cannot even visit project sites to measure river flows, meaning only six months of the year can be used to site projects." The result is variations in performance with users reporting that two-thirds of the microhydro systems installed by the REDP have unplanned voltage fluctuations and outages damaging equipment.[18] Interestingly, some of the units we visited, such as the one at Kavre, had excess electricity during the day but *not enough* during the night, when they had to force a teashop to use less electricity and curtail the use of televisions at some homes so others could have sufficient lighting. Village elders also asked furniture shops to relocate outside of the community since they were consuming too much power.

In terms of maintenance, respondents reported a "lack of trained staff" and "only a few dozen small businesses engaged in maintenance," clearly "an insufficient number for the amount of existing projects." One of the microhydro units we visited in Lukla, for example, broke down less than 2 years after it had been installed and took weeks to be repaired, with a broken runner having to be taken all the way to Kathmandu via airplane to be fixed. As one respondent noted, "the intakes for microhydro plans need continuous daily maintenance as well as seasonal cleaning of sand and stones, yet in some projects this never happens, lowering performance and shortening lifetimes." Another unit we visited near Kavre was "breaking down every three to four months" because it was "old" and "in dire need of an overhaul."

---

17   Mainali, B. and Silveira, S. 2011. Financing Off-grid Rural Electrification: Country Case Nepal. *Energy* 36(4), 2194–2201.

18   Banerjee et al. 2010.

In terms of manufacturing, respondents noted that "companies making microhydro parts are overloaded" and that "a minimum six-month delay" existed for new projects in Nepal. As one respondent explained:

> The problem is both lack of parts and technical manpower. We've been losing engineers to the Gulf countries creating a shortage of labor for welding, electronics, and manufacturing. This was made worse by the civil war. We lack staff and resources, so much that once we build a unit we often have no one to deliver it to the village.

Yet another respondent confirmed that "having consistent manufacturing standards and designs is difficult, given the fragmented nature of microhydro manufacturing in Nepal."

Economic barriers relate to low plant load factors, financing, and tariff collection. The most consistent challenge cited by more than a dozen respondents was the low load factor of microhydro schemes. Because communities think they need only electricity at night, most excess energy is simply "discarded" during the daytime. The average load factors, for example, of the microhydro units we visited varied from 25 to 30 percent, meaning 70 percent of the energy was "wasted." As one respondent put it:

> Most people in rural areas are not savvy; they don't know what to do with electricity beyond lighting a single lamp per household. Teaching them that they can do more with a system—cook with electricity, husk rice, separate mustard, grind corn—is really hard.

Another respondent put things in perspective by noting that most microhydro units produce electricity 24 hours a day, but that communities tend to use that electricity for less than six hours a day. The result is not just lost energy, but lost productivity. Microhydro systems "don't make as much money as they should, so it's harder for communities to pay off their loans. It's a horrible combination of low incomes and low load factors." One study confirmed this in their survey which found that the average microhydro community relied on systems for only 4–5 hours of lighting at night and had load factors below 33 percent.[19]

A second economic challenge is financing. Private banks and vendors are "generally not interested in microhydro." They would rather "loan money to other projects, given the risks perceived in dealing with poor, rural communities." One study found that most interest rates for microhydro loads exceeded 15 percent, "very high" and meaning if turbines break down for even a few days, communities can quickly lose money on a project.[20]

---

19   Banerjee et al. 2010, 17.

20   Pokharel, S. 2003. Promotional Issues on Alternative Energy Technologies in Nepal. *Energy Policy* 31, 307–318.

A third economic challenge encompasses tariff collection, both in setting tariffs too high and too low. The average amount charged per kWh is 1.5 rupees per Watt per month, not enough for many microhydro plants to break even.[21] In other areas respondents noted that "tariffs are not collected at all" or that "electricity use is not metered," making it difficult to calculate who consumes, and who owes, what. As one respondent noted:

> Household metering equipment is generally not installed, meaning households pay according to the number of light bulbs, or social status, or a cutoff device that limits consumption to 100 W. The problem is that some homes claiming to use just 100 W actually have televisions, radios, and DVD players, while others have just one light bulb. These different households, rich and poor, pay the same amount for their very different electricity consumption.

This flies in the face of UNDP guidelines that different tariffs be matched to the socio-economic status of families.[22] We also found great variation in rates between communities, with those in Kavre paying 10 rupees per bulb per house but 100 rupees per bulb per house in Lukla. One respondent noted as well, that in some cases, grid connected electricity, when available, was 50 to 60 percent cheaper than off-grid microhydro electricity.

Political difficulties include institutional challenges and aid dependency. One respondent noted that "the government is too poor to initiate anything" and another that "Government agencies in Nepal, even the AEPC, have their limits. They have systemic human capacity problems, a limited number of professionals." Another noted that "government and donor targets are ambitious, but are lacking capacity to meet them in terms of both manufacturing and human resources." Yet another noted that "the AEPC is understaffed and donor dependent, progress is not as fast as it could be."

Aid dependency means that "different international donors give different priorities, some of them competing and conflicting with the REDP." One respondent argued that:

> A donor dysfunction exists in Nepal where sponsors give money too quickly, but with no sense of responsibility, funding microhydro plants that work for a few years and are then quickly forgotten about. Some of the funds of REDP feel this way, a sense of long-term sustainable operation is not inculcated, they clap and pat themselves on the back after a project is built, feeling good about themselves, and then forget about it.

---

21   Banerjee et al. 2010.

22   United Nations Development Programme 2006. *Vulnerable Community Development Plan (VCDP) Framework*. Kathmandu: UNDP's Renewable Energy Development Program.

As another mused, "Donors are easy with the money, they can't wait to spend it, and they don't even care if it goes to bad projects as long as they can check it off their list." In this sense, the potential for microhydro is so large that "donors can make pretty big mistakes, install slipshod systems, and still see real progress made." A final complication from aid dependence is that microhydro units "must compete with other alternative energy technologies being pushed by donors, such as improved water mills, biogas plants, and solar home systems."

Social challenges, lastly, entail community disagreement, equity, and agricultural processing. Many microhydro systems are shared by multiple communities, but this requires good relationships between upstream and downstream riparian villages.[23] Yet respondents noted some cases where "conflicting schedules for use of the plant," "disagreement over whether to invest or not," "conflicting plans for land use," or "drinking water and irrigation coming before hydro" have all stopped facilities from being built or maintained. One respondent argued that "choosing a microhydro system is a big community decision. There will often be intense discussion and disagreement about whether to proceed, and then which type of system or vendor to rely on."

In addition, the cultural background of rural communities in Nepal is not egalitarian, it is hierarchical and exclusionary. Most rural villages informally adhere to a caste system based loosely on Hinduism that places Brahmins, Chhetries, and Newaris on the top and Damais, Kami, and Sarki on the bottom.[24] Inserting microhydro units into this uneven social system can "exacerbate inequalities," with in some cases "those from the lower ethnic groups contributing labor during construction not given access or an electricity connection." Subsidies from the government can exacerbate this hierarchy when paid to individuals rather than communities.[25]

Finally, some respondents noted that mechanical end-uses, not electricity, are most important for rural communities. One respondent called the REDP program "picking the wrong priority" since "the savings for oil expelling from mustard seed, the most important cash crop, as well as milling and rice hulling, have the broadest social impact." For example, without mechanical energy, women have to get up well before dawn, process daily agricultural requirements, and then cook meals, meaning "electricity doesn't really save them time." Whereas electricity from microhydro units tends to be used by one-third to one-half of all villagers, agricultural processing units tend to be used by all but the poorest 3 to 8 percent of villagers. This means that microhydro units oriented towards processing can create massive social savings, saving 30 to 110 hours for various processing requirements presented in Table 7.2. As one respondent concluded, "common sense for some is nonsense for others. The REDP pr should have focused on agro

---

23   Cromwell 1992.

24   United Nations Development Programme 2010. *Micro Hydro Implementation Guidelines*. Kathmandu: UNDP's Renewable Energy Development Program.

25   Cromwell 1992.

processing first, not electricity." (To be fair, many existing microhydro schemes provide both electricity and mechanical energy for processing, though they do prioritize electricity and new projects are increasingly encouraging electric agro processing).

Table 7.2 **Average time requirements of agro processing in Nepal**

Activity	Traditional (hours)	Hydropowered (hours)
Milling (handmill)	32.3	1.2
Hulling (dhiki)	32.5	1.1
Expelling (khol)	117.5	4.5

Conclusion

Despite these challenges, the REDP expanded microhydro service coverage beyond 30,000 new households and could reach one million people by the end of 2012, ahead of schedule and below cost, all in the midst of a civil war. To be sure, there are elements of the REDP that cannot be replicated outside of Nepal. Broader efforts, such as the national hydropower development plan, water resource development plan, renewable energy perspective plan, and national climate change policy of Nepal, may have also contributed significantly to microhydro power development. But the REDP produced quantifiable benefits that far exceeded costs, confirmed by multiple independent assessments, and channeled resources into community mobilization funds that empowered women and rural enterprises.

In the face of the Maoist insurgency and a host of other problems, program managers could have cancelled the REDP. Instead they acknowledged challenges, restructured the program twice, and refused to abandon it, "funneling money into components that seemed to be working and pulling the plug on those that were not." In the end, as one responded noted, "the REDP was an octopus of a project, but that's the way energy development works—it's messy and sloppy, and if you're afraid of that, you shouldn't be in this business."

Chapter 8

The Energy Services Delivery Project in Sri Lanka

Introduction

At the turn of the millennium, Sri Lanka faced a series of daunting energy security and development challenges. It was primarily a biomass centered energy sector, with 47.4 percent of demand met from fuelwood and dung, 43 percent petroleum, and 9.5 percent hydropower.[1] Seventy percent of households depend on biomass, mostly cooking, and electricity represented only 7 percent of overall energy use. Moreover, 60 percent of household demand for electricity went to one use only, lighting.[2] About half of the population earned less than $2 per day.[3]

To minimize the health implications of households biomass use, diversify the energy sector, and improve incomes for communities, the World Bank and Global Environment Facility initiated the $55.3 million Energy Services Delivery (ESD) Project in 1997. The ESD aggressively promoted solar home systems (SHS), grid-connected microhydro projects (MHPs), off-grid community based village hydroelectric projects (VHPs), wind turbines, demand side management (DSM), and capacity building. Its key objectives were to provide electricity to rural households, strengthen the regulatory environment in favor of energy efficiency, improve private sector performance, and reduce carbon emissions.

The ESD was an unqualified success. It achieved all of its targets below cost and ahead of its determined closing date of 2002, successfully installing 21,000 off-grid SHS, 31 megawatts (MW) of MHPs, 574 kilowatts (kW) of VHPs, and a 3 MW grid-connected wind farm. By the end of 2004, two years after the ESD's close, the Sri Lankan renewable energy industry had more than 40 mini-hydro developers, 10 registered solar companies, 22 registered village hydro developers, and 12 village hydro equipment suppliers compared to less than 3 of each before the ESD began. Roughly three times the ESD's budget, $150 million, was invested in the market from 1998 to 2004. Furthermore, the ESD attracted private sector

1   Sri Lanka Sustainable Energy Authority. 2009. *Sri Lanka Energy Balance 2007: An Analysis of Energy Sector Performance*. Colombo: SEA.

2   Nagendran, Jayantha. 2001. *Sri Lanka Energy Services Delivery Project Credit Program: A Case Study*. Colombo: DFCC Bank, May.

3   Integrated Development Association. 2004. *Energy for Sustainable Development Sri Lanka – A Brief Report with Focus on Renewable Energy and Poverty Reduction*. Kundasala, Sri Lanka: International Network for Sustainable Energy.

developers into the renewable energy sector through public-private partnerships, set national grid interconnection and tariff standards, and instigated the formation of hydro, wind, solar, and energy efficiency industry groups.

In exploring the history, benefits, challenges, and implications of the ESD in Sri Lanka, this chapter is of value to readers for five reasons. First, and most narrowly, past assessments of the ESD have been done by project stakeholders (possibly biasing them) and primarily focused on the contributions of the ESD until 2002 or 2004; no study as of yet has explored the ESD dispassionately and neutrally, and taken into account meaningful changes to the Sri Lankan renewable energy sector since then.

Second, this study delves into how Sri Lankan planners attempted to supply energy services as they dismantled the functions of the welfare state, promoted privatization and restructuring, and emerged from a 26-year old civil war— providing insight for how such tensions can be managed.[4]

Third, a World Bank sponsored review of the ESD concluded that it "can serve as an excellent model for other rural electrification initiatives,"[5] meaning it offers outstanding lessons to those wishing to expand access, electrify rural areas, and reduce energy poverty around the world. The ESD was also the precursor for a large $133.7 million follow-up project called the Renewable Energy for Rural Economic Development project (RERED), illustrating how planners can scale-up energy access programs.

Fourth, the ESD's focus on renewable energy and energy efficiency highlight how low-carbon options can be successfully promoted in an emerging economy.

Fifth, and most broadly, the ESD was the World Bank's first foray into what has now become known as a "market-based renewable energy services provision model," and it thus illuminates how two very prominent global energy actors, the World Bank and the GEF, now design and implement their energy projects.

Description and Background

In 1996, approximately 70 percent of households outside of Colombo and the Western Province in Sri Lanka had yet to be connected to the national grid.[6] Experts estimated that as many as 300,000 rural households were using power from rechargeable car batteries for basic electricity needs such as lighting, operating water pumps, watching television, and charging mobile phones. With energy

---

4   Caron, Cynthia M. 2002. Examining Alternatives: The Energy Services Delivery Project in Sri Lanka. *Energy for Sustainable Development* 61, 38–46.

5   International Resources Group. 2003. *World Bank/Sri Lanka Energy Services Delivery Project Impact Assessment and Lessons Learnt*. International Resources Group.

6   The World Bank 1996. Sri Lanka Energy Services Delivery (ESD) Project. *Report, Energy, The World Bank*. Colombo: The World Bank.

demand growing at eight percent annually,[7] electrification was quickly becoming a controversial political issue. Although the government had been increasing investments in new electricity generation capacity by about 4.5 percent of national GDP per year since the 1990s,[8] increasingly crippling costs of imported petroleum continued to undermine the Ceylon Electricity Board (CEB), the national utility.

At the time, it was estimated that the cost of extending distribution networks averaged $650 per customer.[9] The government's 1991 Rural Electrification Master Plan thus stated that it would only be economically feasible to connect up to 60 percent of villages and 42 percent of rural households, taking between eight to ten years to reach these targets.[10] In the capital city alone, extended power cuts were a daily occurrence, lasting as long as eight hours at a time, and many rural villagers continued to spend hours each day cutting and chopping wood for household energy use. Thus, it seemed that the prospects for the electricity network to be extended to the rest of the population were quite bleak. The government's energy strategy as outlined in the 1994 National Environmental Action Plan was therefore to explore a wider range of energy technologies, extended to include on- and off-grid solutions as well as renewable energy and demand side management (DSM).

The Energy Services Delivery (ESD) project, funded through a World Bank credit line of $22.3 million and a GEF grant of $5.7 million,[11] was conceived as a viable solution to bridge the widening gap resulting from underinvestment in Sri Lanka's power sector. The project ran from 1997 to 2002, and it harnessed the potential of the country's dynamic private sector to complement government efforts to address urgent rural electrification issues. It also stimulated investments in power generation and improved end-user efficiency. In a span of six years, the project successfully installed 21,000 SHS, 350 kilowatts kW of village hydro capacity, 31 MW of grid-connected mini-hydro capacity, and 3 MW of wind capacity[12]—all ahead of schedule and below expected cost.

Although it was initially conceptualized as an experiment, the achievements and potential scalability of the ESD project were so convincing that the World Bank declared it as "a model for other rural electrification initiatives with renewable energy and energy efficiency components".[13] A cursory glance of Sri Lanka's thriving renewable energy market that has developed since certainly

---

7 The World Bank 2004. Project Performance Assessment Report: Sri Lanka Energy Services Delivery Project. *PPAR, Operations Evaluation, The World Bank*, Colombo: The World Bank.

8 The World Bank 2004.

9 Global Environment Facility 1996. Sri Lanka: Enegry Services Delivery (ESD) Project. *Proposal, Global Environment Facility.*

10 Ministry of Power and Energy, Sri Lanka 2011. *Performance 2010 and Programmes 2011.* Colombo: Ministry of Power and Energy.

11 Ministry of Power and Energy, Sri Lanka 2011.

12 Ministry of Power and Energy, Sri Lanka 2011.

13 Ministry of Power and Energy, Sri Lanka 2011.

seems to validate this conclusion. The World Bank continues to apply its market-based approach in similar projects in other developing countries.

To this end, the funding for the project was divided into three principal components, namely:

- The ESD Credit component (estimated at $48.9 million equivalent) that provided medium- to long-term financing for the diffusion of SHS, grid-connected MHP, and off-grid VHP units through local companies, non-government organizations (NGOs) and associations or cooperatives;
- the three MW Pilot Wind Farm component (estimated at $3.8 million equivalent) that was intended to demonstrate the technical and commercial viability of wind power in Sri Lanka in order to encourage future investments by the private sector; and
- the Capacity-Building component (estimated at $2.6 million equivalent) that was a fund to provide training and technical support for renewable energy and energy efficiency initiatives, in particular, for the CEB and energy service entrepreneurs.

Because of this book's focus and methodology, we do not discuss the Pilot Wind Farm or the energy efficiency subcomponent further.

Rather than managing the funds directly, the government appointed DFCC Bank to set up an administrative unit (AU) as a separate entity within the bank, to act as the implementing agency—a formula they continue to use in the RERED project. As part of a blue chip private development bank, the AU was well versed in the prevailing regulations and procedures of Sri Lanka's banking sector. It also took on a quality assurance role by monitoring suppliers' compliance, services standards, and taking in consumer complaints, including tracking over 20,000 SHS. The AU administered the GEF's co-financing grant funds available to off-grid projects to further overcome the initial cost of these technologies, raise awareness within off-grid communities, and facilitate other technical assistance-related activities.

The ESD project intended to encourage a transition in the power generation sector in Sri Lanka that was currently still dominated by an inefficient public sector monopoly to one were the private sector and renewable energy played an increasingly important role. Thus, the largest component and by far the centerpiece of the project was its credit component. This part of the project focused on promoting private sector participation in the diffusion of renewable energy technologies for the purposes of rural electrification as well as additional power generation. Targets encompassed needing to install 30,000 SHS (revised to 15,000 units during the project midterm review in 2000), 250 kW through 20 village hydro systems, and 21 MW of grid-connected mini-hydro systems.

One defining characteristic of the Credit Line was its phased-reduction of GEF grants. Rather than cover costs entirely, the component gave a series of grants on a sliding scale. At the start of the program, all SHS received a 15 to 20 percent subsidy, and off-grid VHPs received a one-time subsidy of $400 per kW. The

GEF, however, also gave performance based grants if costs declined or efficiency improved. SHS dealers received a $2.30 subsidy per Wp for offering smaller sized systems over time. However, these grants were slowly phased out so that by 2002 they covered only 8 to 12 percent for SHS and $200 per kW for VHPs, and by 2004 they did not exist at all.

The credit facility that the World Bank's IDA provided was channeled through the Central Bank of Sri Lanka (CBSL) at concessionary rates. The basic financial model was designed to take into account the fact that financial institutions and the private sector in Sri Lanka were not yet familiar with renewable energy projects and would thus be risk averse when faced with the high capital and transactions costs and the long maturation period for the loans.

Funds were disbursed through a select number of PCIs consisting of private banks, leasing companies, and microfinance institutions that assumed the credit risk of each loan at the average weighted deposit rate (AWDR),[14] repayable in 15 years with a maximum five-year grace period. They in turn offered sub-loans along with their own complementary financing to local companies, NGOs, associations, and households, with a maximum maturity of ten years with a two-year grace period. PCIs were allowed to refinance up to 80 percent of the loan amounts and use their own standard procedures to assess the credit-worthiness of their borrowers (see Table 8.1). "In the beginning, it was all a matter of trial and error," admitted one respondent. "We started with zero knowledge." However, as another respondent put it, "The arrangement gave stakeholders access to long-term financing that was not yet available in Sri Lanka's commercial lending market."

It is helpful to illustrate how the Credit Line worked with a real-world example. At the 12 kW Meddawatte Village Hydro Project in Sabaragamuwa Province, costs were shared almost equally among all stakeholders. The AU provided funds (and covered the risks involved) directly to a PCI, Sampath Bank.[15] That Bank then hired a consultancy company, a material supplier for civil works, and a supplier of electro-mechanical equipment to build the dam and powerhouse, in compliance with local and national standards. All components were made locally except for the turbine shell. The community formed a Village Electricity Society to own and operate the plant, further funded in part with technical assistance from the GEF, with each household giving a financial contribution or contributing labor, land, or other assets. Some, for example, donated up to 300 hours of their time in a practice of *Shramandana* (essentially, "voluntary work in exchange for payment") helping dig a 1 km long canal that connects the powerhouse to the river. Others gave material for civil works, distribution lines, wooden poles for the distribution network, and sand for free. Fifty houses pooled their resources and gave a combined cash payment of 900,000 Rs (~$10,000), about 25 percent of the cost

---

14   Weighted average interest rates paid by all commercial banks on interest-bearing term deposits, as issued weekly by the CBSL.

15   Deheragoda, C.K.M. 2009. Renewable energy development in Sri Lanka: With special reference to small hydropower, *Tech Monitor* November/December, 49–55.

Table 8.1   ESD project lending terms

Technology	Interest rates	Maturity period	Guarantee/Collateral
Grid-connected Mini Hydro	Equal to AWDR plus 4 percent	6–8 (including grace period of 1 to 2 years	Project assets
Village Hydro Systems	Equal to AWDR plus 4 to 6 percent	6–8 years (grace period of 6–12 months)	Project assets
SHSS	Fixed at AWDR	2–4 years (with no grace period)	Project assets and two guarantors from the village

Table 8.2   Stakeholders involved in the 12 kW Meddawatte VHP

Stakeholder	Key actor(s)	Role
Banking Sector	AU (DFCC Bank)	Provided funds to the Participating Credit Institutions, set up procedures to ensure adherence to specified standards
Banking Sector (Participating Credit Institution)	Sampath Bank Limited	Processing loan request, approval and disbursement of loan. Obtaining refinancing from ESD, loan recovery and paying back funds to ESD
Private sector	Consultancy company	Social mobilization, project technical design, assist community in obtaining statutory approvals, environmental clearance, preparation of project feasibility study, assist community in obtaining loan from the PCI, assist community in the procurement process by providing technical specifications, evaluation of quotations, construction supervision and coordinating with the AU
Private sector	Material and equipment supply companies	Supply of pipes, wires and other construction materials
Private sector	Electro-mechanical equipment supplier	Supply and installation of electromechanical equipment
Local Government	Divisional Secretariat	Approval for using the waterway and implementing the project.
Ministry of Environmental Affairs	Central Environmental Authority	Approval for project with conditions for construction of the project with minimum environmental impact.
Community	Meddawatte Village Electricity Board	Owns, operates, and provide after sales services for maintenance of the power plant
Community	Participants	Contributed equity funds, sweat equity, obtained loan under ESD from PCI, manages all funds, participates in the procurement process, obtains all necessary statutory approvals, participates in the construction.

of the entire system. In return, each contributing household will receive a 200 W connection for practically free, paying no tariffs except a small 150 Rs (~$1) monthly fee for maintenance.

By ESD's closing in December 2002, all targets for the credit component had been exceeded well below estimated costs. There were 15 grid-connected mini-hydro systems involving ten private developers, generating 31 MW with more capacity in the planning stages. Four major SHS vendors had successfully installed approximately 21,000 SHS across rural Sri Lanka at a rate of about 1,000 new installations per month. Lastly, 35 village hydro systems serving 1,732 households had also been installed, generating 350 kW (see Table 8.3).

Table 8.3   ESD credit component targets and costs

Project cost	Appraisal target	Appraisal estimate ($m)	Target accomplished	Average unit capacity	Actual cost ($m)
Mini Hydros	21 MW	30.8	31 MW	1 MW	26.7
Solar Home Systems	15,000	14.4	21,000	40 Wp	9.2
Village Hydro Systems	250 kW	0.7	350 kW	12 kW	0.8
Business Development	n/a	0.5	n/a	n/a	1.0
Off-grid Project Support	n/a	1.2	n/a	n/a	0.7
Total		47.6			38.4

SHS Scheme

For the SHS scheme, the original design was for dealers or developers to provide consumer credit on top of handling marketing aspects and providing technical support. This means that the dealers purchase the systems or components directly from the manufacturers by accessing financing from PCIs. The dealers benefit from the low import duties for SHS components reduced from 30 percent to ten percent to further assist with the high upfront costs. They then sell the systems directly to households providing a credit facility together with a subsidy of $100 per system and a co-financing grant from GEF to cover business development, marketing, or capacity building expenses.

However, the SHS dealers soon realized that this microcredit arrangement was too specialized and beyond their expertise. For example, collections were very difficult and time consuming because most of the dealers had yet to develop a strong rural presence. Moreover, because of perceived risks, dealers were reluctant to extend credit to potential customers with little financial history as they had poor local knowledge and understanding of the communities they were serving in the beginning. For similar reasons, PCIs themselves also had very little interest in extending credit directly to potential SHS users. "We do not have the rural networks needed for this kind of financing scheme," admitted one respondent. "These individual loans were too small for us to service," said another.

Thus, in the first three years of the project, the sales for SHSs were quite stagnant, with only about 1,000 units sold by early 2000. This changed with the entry of Sarvodaya Economic Enterprise Development Services, or SEEDS, as a PCI. As Sri Lanka's longest serving microfinance institution, SEEDS had years of experience and grassroots networks in the very rural markets SHSs were being targeted. The entry of SEEDS into the project freed up SHS dealers enabling them to focus on the technical aspects of installing the systems and providing post-sales service, which is their expertise. Moreover, SEEDS used the subsidy from the GEF to incorporate financial incentives for loan officers who earned significantly less than their SHS dealer staff counterparts.

As a result, SHS sales experienced exponential growth that boosted the nascent solar industry, which by the end of the project saw four "strong" solar companies operating nationwide, more than 500 employed technicians, and a combined number of sales surpassing 1,000 units per month. The achievements continued to expand during the RERED project. SEEDS alone, for example, financed more than 60,000 systems during the period 2002 to 2006.

Village Hydro Scheme

The average cost of a village hydro system (such as the one shown in Figure 8.1) was about $2,000 per kW.[16] Under the ESD, villagers contributed up to 25 percent of costs consisting of *shramadana* (manual labor) and ten percent cash. The GEF provided a capital subsidy in the forms of a co-financing grant amounting to $400 per kW and paid project developers staggered fees at predetermined milestones.

In the beginning, PCIs were very skeptical of extending credit for village hydro projects. For until recently, it was illegal for those other than utilities to generate and sell electricity in Sri Lanka. However, an exception was made for Electrical Consumer Societies (ECS) that owned and operated village hydro schemes under the ESD project. Because of the community-centered approach of village hydro schemes, it was not feasible to have individual borrowers responsible for the loan. In order for a village to qualify for a loan, they were required to set up an ECS consisting of villagers within the proposed service area (usually within a two kilometer radius). ECSs were in turn entitled to generate, distribute and consume the electricity produced by members of the society through the village hydro systems.

---

16   Nagendran, J. 2008 *Financing Small Scale Renewable Energy Development in Sri Lanka.* Project Management, DFCC Bank.

Table 8.4   SHSs installed, 1998–2011 (cumulative)

Project	ESD					RERED					RERED-AF			
Year	1998	1999	2000	2001	2002	2003	2004	2005	2006	2007	2008	2009	2010	2011
Capacity (kW)	2	26	109	616	985	1,868	2,904	3,910	4,624	5,170	5,549	5,649	5,722	5,760
Households	50	683	2,574	13,316	20,953	39,530	62,834	83,773	101,551	115,195	124,800	127,560	129,606	130,721

Figure 8.1   The 12 kW village hydro system in Meddewatte, Sri Lanka

To convince PCIs about the credit-worthiness and reliability of the project, the presence of a developer was required. Project developers had to be registered consultants with the AU. Typically, project developers raised awareness in the villages regarding the possibility of building village hydro systems. Once villagers were mobilized, project developers prepared a feasibility report with detailed engineering calculations in accordance with the required technical specifications, assisted the ECS in obtaining all required environmental and statutory clearances, negotiated the loan from a PCI, and provided technical assistance during project implementation. A verification of technical compliance was made at the design stage and again upon project completion by Chartered Engineers, the apex body for the engineering profession in Sri Lanka.

Although it was an unconventional arrangement, village hydro schemes gained credibility. The community-centered approach provided a strong social control for members within each ECS both in terms of electricity usage as well as payments. Out of 35 village hydro systems set up during the ESD project, only three were known to have defaulted and the approach continued to be a success during the RERED project. Moreover, most systems were still in operation at time of writing this book despite some areas already having grid electricity, which indicates the high desirability of the systems. "Usually half the village chooses the grid and the other half sticks to the hydro system," stated one respondent. As Table 8.5 shows, village hydro capacity grew from a meager 22 MW serving 140 households in 2000 to more than 1,900 MW and 7,500 households in 2011.

Table 8.5   Village hydro systems completed, 1998–2011 (cumulative)

Project	ESD					RERED					RERED-AF			
Year	1998	1999	2000	2001	2002	2003	2004	2005	2006	2007	2008	2009	2010	2011
Capacity (kW)	0	22	75	128	350	661	810	1,011	1,171	1,432	1,577	1,737	1,876	1,964
Households	0	140	365	573	1,732	2,548	3,817	4,587	5,129	5,869	6,425	6,803	7,233	7,504

Grid-connected Mini-Hydro Scheme

At the beginning of the ESD project, conventional hydropower was the only indigenous renewable energy source developed in Sri Lanka, providing about half of the country's commercial power needs.[17] The mini hydro industry, however, was virtually non-existent with only a one MW privately owned power plant operational in 1997.[18] To stimulate the interest of the private sector, the ESD project introduced two important innovations. The first was a standardized Small Power Purchase Agreement (SPPA) that covered a period of 15 years on very lenient terms. The second was a Small Power Purchase Tariff (SPPT) mechanism which at the time was based on oil avoidance costs favoring mini-hydro technology. These financial tools allowed small developers (projects of up to 10 MW in capacity) to overcome high transaction costs and their inherently weak bargaining power with CEB.

As a result, about ten private mini-hydro developers were able to install 15 mini-hydro power plants, generating 31 MW–10 MW more than the appraisal target. Moreover, average installation costs of $963.5 per kW were achieved against the appraisal estimates of $1,030 per kW, contributing competitively to the goal of least cost power generation. Under the RERED project, the industry has continued to develop and in 2011 achieved 175 MW installed capacity through 85 projects, and it is operating profitably at an average 40 percent load factor. Going forward, another 82 mini-hydro plants are currently under construction and could possibly contribute an additional 172 MW in capacity in the next few years.

Table 8.6   Grid-connected mini-hydro projects, 1998–2011 (cumulative)

Initiative	ESD					RERED					RERED-AF			
Year	1998	1999	2000	2001	2002	2003	2004	2005	2006	2007	2008	2009	2010	2011
Capacity (kW)	0	0	0	0	31	35	69	83	104	111	130	154	166	168
Projects	0	0	0	0	17	19	31	38	50	52	62	67	70	71

Stakeholder Engagement and Capacity Building

One of the first tasks undertaken by this component was an extensive feasibility study of 1,048 villages to determine possible sites for SHS, MHP, and VHP deployment. It was this collection of initial market surveys and pre-investment studies where planners discovered that end users were willing to pay slightly

---

17   The World Bank 1996. Sri Lanka: Energy Services Delivery (ESD) Project. *Proposal for Review, The World Bank.*

18   The World Bank 1996.

more upfront if energy services were more reliable and safer—e.g., SHS and VHPs were perceived more favorably than kerosene and diesel. The ESD established a Technical Advisory Committee to set standards for manufacturers. It provided funds for the CEB to prepare a National Renewable Energy Strategy and establish a Pre-Electrification Unit within the utility to provide support and training to the Credit Component discussed above. Funds were also available to PCIs to prepare feasibility studies, business plans, and document bank loans, and grants were given to developers and village organizations to raise awareness about the ESD and promote the proper installation of equipment.

Stakeholder engagement and continued capacity building opportunities under the ESD project were certainly key factors for the emergence of Sri Lanka's vibrant renewable energy industry. Many respondents have also credited the government's hands off approach as the main reason for this phenomenon. "By leaving the implementation of this project to the private sector, there has been very little corruption and instead, the many interactions among stakeholders have resulted in plenty of innovations and creative approaches," stated one respondent.

The Sri Lankan Business Development Center (SLBDC) was tasked with the bulk of awareness creation and capacity building activities, especially for off-grid electrification projects (estimated to cost \$2.6 million equivalent). It conducted village-level workshops, marketing campaigns (TV, radio and newspapers), and door-to-door promotional efforts in the early phases of market development until dealers and developers could gain a critical mass of potential customers. Figure 8.2, for example, shows one of the project's promotional posters that relied on the idea that being able to generate their own electricity would be a compelling idea for villagers in light of the CEB's failure to provide such a service.

The ESD project also benefited from the involvement of local governments firstly as interlocutors to the communities that were being targeted and later on as additional funders in off-grid projects, especially for village hydro schemes. Mobilization of stakeholders led to the creation of various specialized associations such as the Solar Energy Industries Association, the Grid Connected Small Power Developers Association, and the Federation of Electricity Consumer Societies, which helped pool collective bargaining power when dealing with the World Bank, the government, and PCIs. In the end, it ushered a shift in mindset amongst stakeholders, "from a dole-out mentality to one that is based on least cost solutions with smart subsidies on a needs only basis.

Benefits

During the six-year period of the project, all targets were met and in most cases, exceeded, ahead of schedule. With project costs amounting to only \$38.4 million equivalent, the ESD project also ended up saving almost \$10 million more than originally estimated (please refer to Table 8.3). Although the bulk of the funding

Figure 8.2   Poster promoting microhydro in rural Sri Lanka

came through IDA and GEF's credit line and grant facility respectively, PCIs, the private sector and end-users provided a significant amount of complementary funding, further demonstrating the commercial viability of the approach.

Perhaps the clearest indication of the project's success is that almost a decade after its completion, the successor RERED project, which concluded in 2011, continued to build upon the earlier achievements of the ESD project. In addition, the government has also targeted to increase power generation from non-conventional renewable energy sources to ten percent by 2015.[19]

Apart from these overall successes, the ESD project yielded real benefits for stakeholders on the ground. The following section elaborates further on the positive changes that the project brought about in the communities targeted as well as in catalyzing the development of Sri Lanka's renewable energy industry.

End-users

The ESD project allowed villagers to take charge of their electricity needs in the absence of government provision, and potent participation from local banks and villages in the ESD is a testament to its success. Over the course of the project, PCIs contributed $16 million of their own funds, several NGOs were operating in tandem with the ESD project areas, and more than 30 villages were asking for renewable energy systems. As one bank manager told us, "we weren't participating out of any sense of charity, we were driven by commercial viability. My own bank has lent to 65 various hydro projects and have only had one bad loan." Members of the AU also told the research team that "no major complaints"

---

19   Ministry of Power and Energy, Sri Lanka 2011. *Performance 2010 and Programmes 2011.* Colombo: Ministry of Power and Energy.

had ever arisen regarding the SHS and grid-connected MHP components, and only "one complaint" was registered for the off-grid VHP component—and that was "quickly resolved."

Indeed, an extensive survey done by the World Bank at the close of the project found that energy access through the ESD promoted income generating activities, increased safety, facilitated longer studying, and motivated a variety of other activities shown in Figure 8.3. SHS units, for example, cost only $750 over ten years, including battery replacement, but many rural homes spend $650 per year on kerosene and automotive batteries—meaning units pay for themselves quite quickly. Same with off-grid VHPs: a community of 120 households spends about $27,000 per year on kerosene and batteries, but a VHP system can cost as low as $33,000.

Many of the benefits to the ESD project were confirmed by an independent, follow-up study by the United Nations Development Program.[20] This study found that off-grid VHP units reduce household kerosene consumption by 4 liters per month and completely eliminated the need for battery charging; SHS reduced kerosene consumption even more from 11 liters to 0.7 liters, and eliminated expenditures on charging batteries in 93 percent of households. These numbers add up: about 41.2 million liters of kerosene was displaced from SHS and VHP systems in 2008, and grid-connected MHPs saved 1.2 million tons of carbon dioxide due to their displacement of oil-fired generators. Moreover, 50 to 90 percent of households surveyed stated that access to electricity from SHS and VHP provided better lighting, led to longer studying hours, cultivated a greater sense of safety and security, and facilitated the introduction of televisions, electric irons, radios and mobile phones. Sixty to 90 percent of VHP households specifically reported "increased unity" among villages as a result of "building and managing the subprojects together."

Apart from the electrification, villagers also benefited from being trained in the proper maintenance of their systems. The ESD project improved finances of villagers as there were no monthly payments needed for SHS and hydro systems after the loans were paid off. Having electricity encouraged productive activities such as sewing and carpentry which was further expanded under the RERED project. Villagers also cited health improvements as one of the main benefits of the project as they no longer had to use kerosene. The availability of TV was something that was specifically mentioned and, indeed, during our site visits we observed that black and white TVs were available in all the households visited and in schools, like the one shown in Figure 8.4. However, it was the possibility of studying at night which seemed to be the most important benefit of electrification for the villagers. "Education is very important for Sri Lankans," claimed a respondent. "Even villagers want their children to be highly educated. They do not want their children to become farmers like themselves."

---

20 United Nations Development Program, *Renewable Energy Sector Development: A Decade of Promoting Renewable Energy Technologies in Sri Lanka*. Bangkok: UNDP Regional Center, January, 2012.

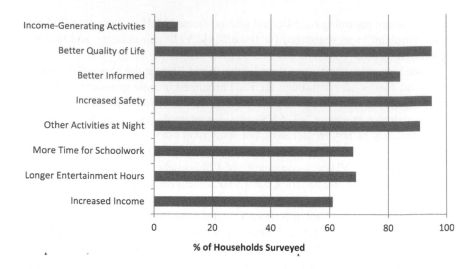

Figure 8.3   Social impacts of ESD project on rural households in Sri Lanka

The project was certainly an opportunity for commercial banks to venture into development infrastructure. Capacity building for small private sector companies and NGOs resulted in many local innovations developed regarding quality standards, microfinancing schemes, and marketing opportunities. It also brought in new players into the industry with additional working capital loans for expansion and training. Most impressively, Sri Lanka's experience and expertise, particularly in mini-hydro project development and financing have been sought out by other developing countries keen to develop their own renewable energy sectors, particularly in African countries such as Rwanda and Uganda. According to one respondent, "The government was able to step back and the private sector just ran with it. The results have been superb."

Lastly, the ESD saw the price of renewable technology drop for Sri Lanka as a whole. The program saw the costs of SHS decline, from $11 per Wp in 1998 to slightly less than $10 per Wp in 2002. The installed cost of grid-connected MHPs declined slightly, dropping from about $1,030 per kW to less than $970 per kW, as well as off-grid VHP prices, sliding from $2,060 per installed kW to $2,020 per installed kW.

Capacity Building

Though more difficult to measure, the ESD strengthened capacity in a variety of ways. For the SHS subcomponent, the ESD supported the design and enforcement of technical specifications for SHS and also trained technicians to conduct spot checks, develop after sales service models, and provide customers a way to lodge complaints.

It helped the grid-connected MHP subcomponent by enabling manufacturers, developers, and PCIs to improve their accounting practices, record keeping, reporting, evaluation, and monitoring. It sent some industry representatives overseas to learn about microhydro design by European firms, including a one month residential program in Germany.

It similarly improved the off-grid VHP sector, creating technical standards and requiring that all designs pass an inspection by Chartered Engineers (a respectable national group) and the Sri Lanka Institute of Engineers. It assisted villagers in hiring consultants to help them with project preparation, and offered classroom sessions on topics ranging from hydrology to maintenance—all paid for with grants, rather than loans, which averaged $6,000 to $8,000 per village. NGOs such as the Intermediate Technology Development Group (later renamed Practical Action) and the Sri Lanka Business Development Council utilized funds to train 30 village cooperatives in social mobilization and hydro development. Capacity building efforts further reduced the turnaround time for loans—which took more than 90 days when the ESD began, but were reduced to less than 30 days by 2001.

Part of what drove such high sales and satisfaction levels in all three renewable

energy sectors were new sales techniques developed and perfected with technical assistance from the ESD. One was displaying products to large groups of people rather than individuals; vendors and PCIs sent speakers with units to garment factories, schools, and hospitals, giving demonstrations and/or answering questions during lunch breaks or between shifts. Another was innovative displays: as one respondent put it, "our rural salespeople went and fixed SHS in temples, churches, and community centers so people could literally see what they could do." Another was door to door visits for SHS done at night, so people could see firsthand what rechargeable torches and electric lights can offer them. A fourth was dealers traveling by motorbike to loan sample systems for a single night so households could become familiar with solar electricity. A fifth was targeting women as beneficiaries since

Figure 8.4   A girl watching an educational program on television powered by microhydro electricity at a rural school, Sri Lanka

household surveys revealed that electricity access benefitted them the most—as it enabled labor saving appliances, reduced household chores, provided access to entertainment, and positively impacted family routines. A sixth was sending along bank officers with the technicians doing demonstrations so interested community members could sign up on the spot.

The ESD lastly promoted renewable energy more generally by contributing to awareness raising. One aspect was the creation of the Sri Lanka Energy Forum, which ran a nationwide awareness program about the ESD. It especially targeted provincial officials and village decision makers and trained them in basic renewable energy concepts, financial options, and feasibility studies; once they were ready to consider systems, it facilitated workshops on appropriate designs. It also sponsored classes at the National Engineering and Research Institute and local technical colleges. It provided assistance that helped create the Federation of Electricity Consumer Societies, an organization that includes all village societies managing off-grid VHPs so that grievances could be represented and the World Bank approached if problems arose; it also served as a platform to share best practices about operations and maintenance. This Federation ended up banning poorly performing companies and a few VECs from participating in the ESD. A final track promoted a national advertising campaign entitled *Gamata Light* run by the Sri Lanka Business Development Centre which featured newspaper ads, television ads, and radio broadcasts promoting SHS and off-grid VHPs.

These awareness efforts were not "one way." The ESD sought feedback from households, VECs, and operators themselves through formal surveys and informal workshops. Based on this feedback, planners learned that common problems facing the SHS subcomponent were improper battery charging, wiring defects, loose connections, and incorrect mounting of the solar panel. For VHPs and MHPs, common difficulties included low voltage, turbine breakdowns, lack of water, and flickering light bulbs. As both of these sets of problems were caused by over usage, the ESD directed education and awareness programs to focus on optimal load patterns.

Challenges

Having just emerged from a protracted civil war, Sri Lanka is now entering a new phase in its history. The country overwhelmingly reelected President Mahinda Rajapaksa in the 2010 general elections, handing over a strong mandate to heal the country, rebuild the economy, and accelerate the pace of development. Sri Lanka's nascent renewable energy sector may therefore have to contend with a new set of challenges.

Renewed Commitment to Fossil Fuels

It is clear that the government's main goal seems to be to quickly restore development particularly in the Northern and Eastern parts of the country that were ravaged by the war and kick start Sri Lanka's industrial development, including investing in energy infrastructure. Figure 8.5, for instance, shows a mini-hydro station at the Carolina Estate abandoned during the civil war, its powerhouse and electrical machinery completely gone. The construction of two new major coal power plants are currently in progress in Puttalam, Northwestern Province (900 MW), and in Trincomalee, Eastern Province (1,000 MW), which are seen as harbingers of economic growth for the country. In fact, these plants have become such potent national symbols that Sri Lanka now lists them on their currency. Moreover, with 88 percent of households electrified thus far and ambitious targets to achieve 100 percent electrification by 2012,[21] energy access no longer seems to be an urgent priority.

Low Capacity Factors

Microhydro units have experienced "far lower" capacity factors than expected. This is partly because of the weather, maintenance difficulties, and changing personnel. One study noted that VHPs and MHPs were "vulnerable to seasonal fluctuations in water flow," especially alterations in the timing of monsoons.[22] One of the operators of a grid-connected MHP we visited explained a "big difference in output" due to "maintenance concerns" as well as "unpredictable weather events." A CEB manager also told the authors that MHPs are essentially "non-dispatchable" since they cannot store their water in reservoirs. In addition to weather and maintenance, a third constraint is "when those trained at each village die or simply move on to other jobs, leaving no one left with knowledge about how to care for units."

Market Saturation

Though the ESD was an undeniable success when it closed, changes since then have mitigated some of its progress. Within the Credit Component, SHS sales "are practically zero" as of 2012 due in part to the grid-electrification plans described above as well as the bankruptcy or closure of three of the major four SHS distributors, and a reduction of operating companies from 15 at the height of the ESD to 2 today. As one respondent noted:

---

21   Ministry of Power and Energy, Sri Lanka 2011. *Performance 2010 and Programmes 2011.* Colombo: Ministry of Power and Energy.

22   World Bank. 2003. *Implementation Completion Report for an Energy Services Delivery Project.* Washington, DC: World Bank, June 4, Report No. 25907, pp. 25–26.

Figure 8.5   An abandoned 2.5 MW grid-connected mini-hydro system at the Carolina Tea Estate, Sri Lanka

> The situation with solar is not so good, frankly. Solar vendors have largely shut down their offices and left customers who need maintenance or repairs in a lurch. Financial institutions have mostly stopped lending due to higher default rates than expected; I've been told that they could be as high as 30 to 40 percent in some areas. SHS are now only sold in the North or East, and then on supplier credit rather than personal guarantees, which had been the norm.

Though numbers could not be confirmed, one respondent claimed that "30 percent of the SHS units sold through the ESD were no longer working" as of 2010, and that at least 7,000 units had been sent back to PCIs for various reasons. A post-ESD 2005 assessment of the Sri Lanka solar industry also warned that gross and net profit margins were being reduced due to higher supply costs and increasing competition, that consumer finance was not available in a quantity sufficient to continue growth, and that survey findings indicated low to moderate levels of customer satisfaction with technical performance.[23] A respondent from one of the PCIs also told us that in 2009 the default rate for SHS was above 50 percent, convincing managers there to quit lending for that product. Another contributing factor was supplier specific units that required proprietary maintenance when things went bad—in some cases beyond the financial means of clients, in other cases repairs became impossible when that supplier was no longer in business.

---

23   James R. Finucane, *Solar Industry Growth Analysis: Sri Lanka.* Colombo: DFCC Bank, June, 2005.

Community Disagreement

At an entirely different scale, community disagreement concerning siting, tariffs, access, and priorities impacted ESD projects. One former manager told the research team that even though there were few formal complaints, "local opposition was a factor, with communities often disagreeing about where a project should be located, or conflicts between upstream and downstream villages, or worries about the mishandling of funds."

Tariffs for were another contentious point, with communities "unsure of whether to charge a flat rate or a tiered rate based on income."

The distribution of energy was a third factor, as "most microhydro projects did not provide enough electricity for the entire village, forcing leaders to make hard calls about where it would go, and disgruntled villagers sometimes tried to sabotage projects." At one of the off-grid VHPs visited by the research team, the village electricity manager told us that "we're happy with the scheme, but we need more connections ... we're saturated and cannot expand even though we have more and more families that need electricity." Another community leader told us that "the standard 200 W per household is not enough, people need more, and Sri Lankans are highly literate and well educated, meaning even rural ones know precisely which electric appliances and devices they want if they happen to receive electricity." (As an aside, the official literacy rate for the country is well above 90 percent).

A final dimension of this challenge relates to gender and priorities. One local consultant told us that men, who often make decisions for communities, will prioritize energy access for themselves and their own homes even if it means excluding others; will put that energy to use for irrigation; and prioritize information from radios about crops and agriculture. Women, by contrast, prioritize energy access for everybody; want energy for cooking and water pumping; and seek entertainment and lighting as the most desired services. In essence, "Sri Lankan men and women have completely different ideas about the priorities of off-grid electrification."

Poverty

Although the ESD certainly tried to help the poor, respondents commented that "it was really private sector driven, not poverty driven" and "the prime motive was to build industry, not help poor households." Another respondent cautioned that "on paper, the ESD project did a lot of good. However, not enough is being done for poorer households. There should be schemes that cover them too."

In theory, the involvement of SEEDS and microfinance institutions—the so called "barefoot banking" where loan officers personally know and understand those they lend to—was supposed to ensure poor communities profited. But in practice, part of the challenge appears to be that "most of the PCIs were comfortable working with rural companies and enterprises, but not rural households."

Another part is that commercial firms had stronger political connections. One respondent argued that:

> Suppliers, rather than the needs of communities, villagers, or even banks, drove the ESD. This is why it promoted SHS—lots of suppliers there—instead of solar drying of foods, solar cookers, cookstoves, or solar lanterns. The SHS lobby was almost like a 'mafia' in how they influenced the design of the ESD.

A second respondent argued that "The ESD favored SHS not based on merit, but on industry influence."

A third explanation is that the biggest beneficiaries of the ESD appear to be middle and upper class rural homes rather than those in the lowest economic quintiles. As one respondent noted, "the ESD's benefits are skewed towards higher class families who have better willingness and ability to pay; most of the poor have only seasonal income and were too risky for PCIs." These wealthier families had either land themselves or regular incomes from tea production or small-scale agriculture, compared to poorer families who lacked both land and income—making them a hard demographic to loan to. A few respondents mentioned the adage that "bankers tend to prefer to lend to those that don't need their money." As a result, "the poorest were untouched by the ESD."

Consequently, vendor bias towards wealthier clients manifested. One respondent noted that "SHS sales were partly commission based, meaning dealers didn't want to bother with smaller, less expensive units and poorer households." Solar companies matched bonuses and incentives with sales revenue as well, further creating momentum to sell larger units—one company even gave their top salespersons a free automobile. These dealers "were in the solar game to make money, pure and simple, not to address poverty; I bet if most of these salespersons bumped into their customers on the street, they wouldn't even recognize them—it was all about profits, not people."

The idea that the ESD did not truly raise incomes and reduce poverty has been supported by a few independent assessments. One from the University of California Berkeley argued that the ESD did not deliver the rural economic benefits it was supposed to[24]. It was especially critical of the SHS component, noting that only 8 percent of households that purchased systems apparently reported direct economic benefits. It concluded that rather than giving 85 percent of its off-grid funds to SHS, it should have given the majority of funds to VHPs, though these were only "slightly better" with 20 percent of recipients reporting direct economic benefits. The study concluded that instead, the ESD should have prioritized VHP connections to schools, health centers, and small enterprises and solar water pumps. Others have suggested that cookstoves would have represented sounder

---

24   Kapadia, Kamal. 2003. *The Not-So-Sunny Side of Solar Energy Markets: A Case Study of Sri Lanka.* Berkeley: Energy and Resources Group at the University of California Berkeley, Master's Project, May 20."

off-grid investments compared to SHS and VHPs.[25] Many of these concerns seem validated by the fact that the successor to the ESD, RERED, explicitly (a) promotes the distribution of electricity services to the poorest households and (b) has incentives for community connections to schools, health clinics, enterprises, and for street lighting.

Socio-Cultural Attitudes

One final challenge concerned socio-cultural views of SHS units, which did not necessarily fit into the rural lifestyles of villagers. Respondents told us, for example, that most rural homes are made of wattle and daub with *illuk* grass for roofing, often collected from small scale rice cropping. Some villagers reported believing that SHS, by providing electricity, could catch those roofs on fire, with one explaining that a solar unit "could catch his house on fire, just like lightning." Others indicated that they believed they would have to upgrade their roofs from coconut fronds and *illuk* grass to ceramic tiles—the same practice CEB requires for grid connections—in order to install SHS, which in actuality need only a roof mounted pole such as the one shown in Figure 8.6 and no other upgrades.

Conclusion

As a largely successful endeavor, the ESD project demonstrates the viability of improving renewable energy technology alongside private sector competence. The building blocks for its success, namely, being able to seize the right opportunity, a well-designed financial model and credit facility, a committed and competent implementing agency, as well as stakeholder participation and capacity building, have resulted in many innovations that continue to reward stakeholders and drive new developments in the renewable energy sector. Moreover, these factors are also part of the success stories of many similar renewable energy projects funded by the World Bank, including its successor RERED project, and initiatives in Bangladesh, the Philippines, and elsewhere.

Although initiatives like the ESD and RERED projects continue to support the development of renewable energy technologies, most investments seem now to be concentrated toward large-scale infrastructure projects. Additionally, as the country prepares for 100 percent electrification by 2012, innovative off-grid solutions have lost much of their original allure. Already, the SHS industry is suffering from market saturation and if new business models are not developed, there is a danger that other renewable electricity technologies will follow suit.

---

25 Wickramasinghe, Anoja. 2008. *Issues of Affordability and Modern Energy Technology: Barriers to the Adoption of Cleaner Fuels for Cooking.* Colombo: National Network on Gender and Energy.

Figure 8.6   A solar panel connected to a SHS in the Western Province of Sri Lanka

Chapter 9
The Village Energy Security Programme in India

Introduction

India is a country that is still overwhelmingly rural, with approximately 70 percent of its 1.2 billion people, or close to 800 million, living in rural areas—the largest in the world. For this reason, India's economic and social development is inherently linked to growth in the rural sector and access to modern electricity and fuel sources. Despite more than a half-century's worth of government efforts to improve rural electricity infrastructure, household electrification levels and electricity availability is still far below the world average.[1]

To put it differently, as of 2009, 404 million Indians reside without electricity. Furthermore, rural Indians lack access to modern commercial fuels for activities such as cooking and heating. A majority of rural households depend on traditional biomass to meet their domestic energy requirements. Even where rural communities have access to LPG, cost is a significant barrier; because people can gather fuel wood for free, they are unwilling to expend the cash necessary to purchase LPG.

This chapter presents a detailed analysis of the Village Energy Security Programme (VESP), a project dedicated to increase energy access for India's rural population through the use of community-scale biogas units. It explores the VESP's project objectives, examines the technologies harnessed for the VESP, and explains its service delivery model. The chapter then concludes by critiquing the challenges and best practices of the VESP and reviewing the project's structure, technological performance, financial strategy, and feedback from stakeholders.

Although the government designed the VESP in a very novel way, many of the test projects took a long time to implement, and once implemented, less than half of them remained functional. The VESP has since been discontinued and no new projects have been approved since 2009. Nevertheless, several researchers and policy think tanks have argued that programs such as the VESP, which decentralize

---

1 Historically, the government measured rural electrification levels as a percentage of villages connected to the grid, irrespective of how many households within the geographical boundaries of each village were electrified. The green revolution in agriculture was arguably the main driver for grid expansion and reflected the way in which rural electrification was calculated. However, in 2004, the government adopted a new definition for rural electrification and many villages that were previously considered electrified now fall by definition into the un-electrified category.

energy generation using locally and widely available renewable energy sources, are key to enhancing power supply in rural areas and to extending electrification to remote areas in India.[2] Moreover, we find that programs like the VESP that address "total energy needs"—namely, electrification and access to modern fuels—could facilitate lasting economic and social development in village communities, if done properly.

Description and Background

The pace of electrification in rural India has been somewhat sporadic. Though most rural villages (about 93 percent) have electricity access, the electrification rate among actual households is much lower (about 60 percent). Even where electricity access is available, the quality of supply remains poor because power is often unavailable during the evening hours when people need it the most. Only seven states have achieved 100 percent village electrification rates, and five of these states have smaller geographic sizes, making electrification easier. Some of the larger states—including Assam, Bihar, Jharkhand, Orissa, Rajasthan, and Uttar Pradesh—have lagged in terms of their rural electrification efforts.

Though Andhra Pradesh and Tamil Nadu have "officially" achieved complete village electrification, a recent field study undertaken by the research team indicates that many hamlets and forest fringe villages do not have access to any form of electricity, on-grid or off-grid.[3] Thus, there is generally a geographic and an income-based divide in terms of electricity access. Urban areas or upper-income households consume more electricity than do rural areas or lower-income households.[4] Even among urban and rural households with comparable incomes, the former consume much more electricity than the latter. Generally speaking, however, electricity consumption per capita increases with higher levels of income. Low-income groups appear to use electricity mostly for lighting, whereas elevated electricity consumption among upper-income groups can be attributed to appliance use.

---

2   Banerjee, R. 2006. Comparison of Options for Distributed Generation. *Energy Policy*, 34, 101–11; Buragohain, B., Mahanta, P. and, Moholkar, V.S. 2010. Biomass Gasification for Decentralized Power Generation: The Indian perspective. *Renewable and Sustainable Energy Reviews*, 14, 73–92; and Ghosh, D., Sagar, A., and Kishore, V.V.N. 2006. Scaling up Biomass Gasifier Use: An Application-specific Approach. *Energy Policy* 34, 1566–1582.

3   Palit, D. and Chaurey, A. 2011. Off-grid Rural Electrification Experiences from South Asia: Status and Best Practices. *Energy for Sustainable Development.*

4   Pachauri S. 2007. Global Development and Energy Inequality Options. Available at: http://www.iiasa.ac.at/Admin/INF/OPT/Winter07/opt-07wint.pdf [accessed: November 20, 2011].

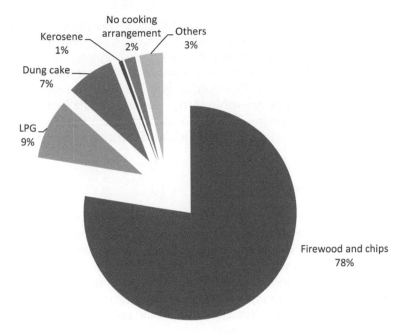

Figure 9.1   Cooking energy use for rural households in India

There is likewise a division between urban and rural areas in terms of access to commercial fuels. Though commercial fuels constitute about two-thirds of total primary energy consumption throughout all of India, the inverse is true in rural areas, where much of the population still has constricted access. In 2007 to 2008, the National Sample Survey Organization conducted a Household Consumer Expenditure Survey and found that more than three-quarters of rural households use firewood and chips as their primary cooking fuel (see Figure 9.1). The proportion of households using firewood and chips is highest among the bottom of pyramid population, namely, 84 percent of agricultural labor households and 90 percent of indigenous or tribal households. Furthermore, about 61 million rural households use kerosene for lighting.[5] Due to a lack of transportation infrastructure, kerosene is less available in the hilly and remote areas of the country.[6]

The government has taken steps to address many of these issues. In 2001, it created the Rural Electricity Supply Technology Mission with the objective of obtaining "power for all by 2012." In April 2005, it continued this effort by

5   64th Round of National Sample Survey, Government of India.

6   Morris, S., Pandey, A., and Barua S.K. 2006. A Study on Kerosene Distribution and Related Subsidy Administration and Generation and Assessment of Options for Improvement of the System. *Final Report Submitted to the Petroleum Federation of India, New Delhi*. Indian Institute of Management, Ahemedabad.

launching the Rajiv Gandhi Grameen Vidyutikaran Yojana (RGGVY), a large-scale program designed to accelerate rural electrification and provide electricity access to all Indian households. Under RGGVY, the government subsidized grid extension to all but the most remote areas. Also, since 2001, the Ministry of New and Renewable Energy (MNRE) expanded plans to electrify remote off-grid areas with locally available renewable energy sources with the Remote Village Electrification (RVE) program. In 2004, it launched the VESP to meet each village's complete energy requirements—cooking, electricity, and motive power.

Description and Background

The MNRE designed the VESP as a scheme that would go beyond rural electrification to achieve "village energization." Resting on the principles of community ownership and locally available resources, the project took a more systematic approach to energy security than its predecessors, focusing on providing clean cooking technologies, using energy for productive purposes, and sustaining local markets. It aimed to support biomass gasifiers, biogas systems, and other hybrid systems, to generate electricity for domestic and productive use, and to disseminate more efficient cookstoves.

Through the program, Village Energy Plans (VEPs) were developed through a participatory approach, with communities and institutions working together to establish projects, and build and manage energy plantations to provide a sustainable fuel supply, either in the form of biomass or oilseeds for biofuel. In a broader sense, the VESP's primary objectives were to:

- create appropriate institutional mechanisms that could facilitate rural energy interventions at a decentralized level;
- strengthen the capacities of different stakeholders to manage the local resources in a sustainable manner;
- meet the comprehensive energy requirements of villages, including cooking, lighting and productive uses; and
- establish a dedicated tree plantation and management system as a feedstock for the village-scale energy production systems.

By delivering comprehensive energy services to rural villages, the program was envisioned to facilitate broader social and economic development in the rural sector by reducing poverty, improving public health, reducing drudgery (particularly for rural women), raising agricultural productivity, creating employment, generating income, and reducing migration. It represented an innovative and pragmatic approach to solving an emerging 3E-trilemma of maintaining energy resources, sustaining economic development, and preventing environmental degradation.

The first phase of the VESP was launched in April 2004. Targeting villages or hamlets in remote rural areas that were unlikely to be provided grid electricity in the near future, the program offered one-time grants (up to 90 percent of total project costs) to Village Energy Committees (VECs), consisting of representatives from villagers and the local governance body (the *gram panchayat*), to install systems capable of meeting energy demands in their respective communities. The VEC usually consists of nine to 13 members, with women representing 50 percent of the committee, and an elected *Panchayat* member from the village serving in an *ex officio* capacity. Villagers were to plan, implement, and sustain their projects with support from external institutions known as Project Implementing Agencies (PIAs). They were also expected to provide equity contributions in cash or in kind.

Furthermore, each VEC created a Village Energy Fund to sustain the project's operation and management. Thereafter, the VEC deposited monthly user charges in this account. The Fund was managed by the VEC with two signatories nominated by the committee. In short, the VEC was responsible for producing the power, distributing the electricity, managing project revenues, and resolving disputes in the event of supply disruptions, an organizational structure depicted by Figure 9.2. The summary of transactions between the various key stakeholders under the VEC service delivery model is presented in Table 9.1.

Table 9.1   Summary of transactions between key stakeholders in the VEC model

Entities	Offers	To	Expects in return	Instrument
Beneficiaries	Fuel	VEC	Payment for fuel supply	-
VEC	Electricity	Beneficiaries	Payment of electricity	Negotiated tariff
Beneficiaries	Payment	VEC	Reliable Electricity	Connection Agreement
OEM	AMC	VEC	Payment	AMC agreement

The MNRE's numbers suggest that the VESP did not achieve its goals, however. As of 30 January 2011, the Ministry reports that a total of 79 VESP test projects have been sanctioned in nine states in India. Of these, 61 projects have been actually constructed or commissioned in eight states, while the others were at various stages of implementation or have not been commissioned. In all, around 700 kW of electricity generation equipment has been installed. Nearly 90 percent of the energy production systems from these projects were based on biomass gasification technology, whereas the remaining ten percent are based on straight vegetable oil engines. Nonetheless, only 42 percent of the 50 surveyed projects were fully or partially operational as of 2011, while the rest were either non-functional (26 percent) or un-commissioned (32 percent) (see Table 9.2).

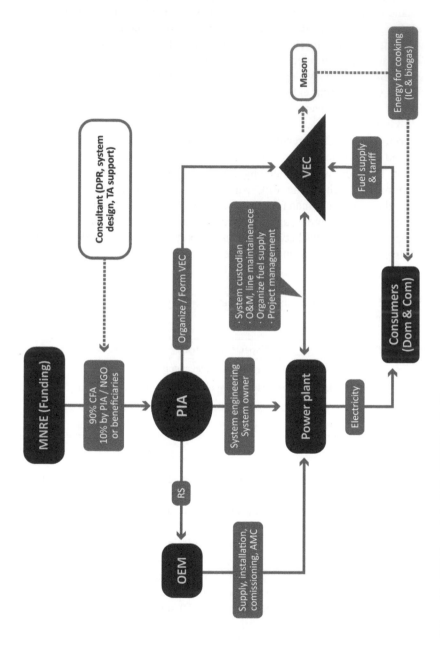

Figure 9.2 The VEC model of the VESP in India

Table 9.2   Status of VESP projects surveyed

Region	No of projects	No of commissioned projects	No of functional subprojects
Assam	14	-	-
Chhattisgarh	6	6	3
Gujarat	2	2	-
Madhya Pradesh	10	9	4
Maharashtra	5	5	5
Orissa	9	9	7
West Bengal	4	3	2
Total	50	34	21

Benefits

Notwithstanding these challenges, the VESP did achieve a collection of multistakeholder, technological, and financial benefits summarized in the following section.

Multistakeholder Participation and Capacity Building

We found that VESP projects emerged as a vehicle to motivate the community, especially the youth, to develop skills in renewable energy development. In almost all projects, local youths learned how to operate the installed energy production systems. In some projects, diesel engine mechanics operating in the neighborhood had opportunities to enhance their skills and enter into annual maintenance contracts (AMCs) with the VEC and the PIA to provide technical support for post-installation maintenance. The innovation adopted by selected PIAs for capacity building of system operators also facilitated improved project performance.

Our findings also indicate that NGOs were better able to mobilize and lead test projects compared with projects implemented by governmental entities. This relative improvement in project implementation was due to close coordination between the NGOs and the VECs. Another reason that NGOs were more successful at mobilizing community support may be that NGO-run projects benefited from the long-term association between NGOs and the villages even prior to initiation of the VESP.

Technological Performance

The VESP did provide a platform for various designers and manufacturers to test new technologies. The small capacity gasifier systems designed by TERI in New Delhi, the Indian Institute of Science in Bangalore, and Ankur Scientific Energy Technologies in Baroda, for example, received opportunities for extensive field testing, allowing these institutions to further customize their technologies for rural markets.

Some villages found innovative uses for their biogas technology apart from as a cleaner cooking option. The community in Mokyachapara village in Maharashtra operates six floating drum biogas plants (with six cubic meters of capacity each) on a communal basis, with each of the biogas plants supplying cooking fuel to five to six households. De-oiled cakes obtained as by-product from oil expellers are also used as feed material for the biogas plant and are mixed along with cow dung to obtain higher gas yield. In Kumhedin village in Madhya Pradesh, 14 domestic-sized biogas plants (each two cubic meters of capacity) are in operation. Using cow dung as feed material, each biogas plant serves one family. Beneficiaries use the sludge produced by the biogas plants as fertilizer for growing vegetables. The vegetables are then sold in the nearby block town to earn additional income. In Amabahar village in Chhattisgarh, a mantle-based lighting point has been connected to each of the biogas plants and beneficiaries enjoy direct illumination from biogas.

Plantations and Fuel Supply Management

We observed that the success of a plantation is proportional to the level of community involvement. In Mankadiatala and Champapadar villages in Orissa, where the Forest Department implements VESP projects, scientific species selection and plantation management practices resulted in survival rates of about 50 to 60 percent—a positive result. The communities planted Simaruba, Bakain, Eucalyptus, Mahaneem, and Acacia trees, which grow well in the region. The community was fully involved in pit digging, nursery raising, sapling planting, and weeding, and with their involvement the Forestry Department planted 1,400 to 1,500 plants per hectare as a block plantation.

In most of these projects, fuel supply is based on each household bringing in a certain quantity of biomass from the community forests to the plant site every week. During the field visits, we observed that the fuel supply was well streamlined in villages such as Dicholi, Karrodoba, Bhalupani and Mahishakeda. In Dicholi and Bhalupani, each household contributes between 30 and 40 kg of fuelwood on a monthly basis. This is collected from fallen wood in surrounding forest areas and within the existing plantations of the village. Electricity payments are also linked with biomass contribution, with non-contributing households required to pay amounts equivalent to what they pay for biomass fuel. In Karrudoba, the fuel supply is mainly provided by the Forest Department from "lops and tops," a term denoting removal of dead branches and dying bark from a tree or shrub. In addition, villagers also contribute biomass from "kitchen surplus," that is, savings achieved because of the use of ICSs.

Financial Performance

The potential for income generation activities exists in almost all VESP communities. The only reason that these activities have not been exploited is because of improper or nonexistent guidance from the VECs. Active involvement of the *gram panchayat* (the lead representative in VEC) tended to help in developing the required synergy between village development funds and VESP funding for initial project costs and operational expenses. For instance, test projects at Dicholi, Bhalupani, and Karrudoba villages swap use of funds from local *panchayat*, state government and local forest offices respectively, either as a capital contribution towards the project or for establishing income generation activities. For example, the local *panchayat* contributed to complete a power distribution line in Dicholi, while the forest department in Karrudoba installed an open well for water supply to the gasifier plant. In Bhalupani, we observed that a honey-processing unit, set up with support from the Integrated Tribal Development Agency, was managed by the Women Federation of the Bhalupani *gram panchayat* in order to generate cash income.

However, based on the field assessment of the VESP test projects, we found that most of the biogas plants have been operating at a CUF of only seven percent. The real challenge is therefore to set a domestic tariff consistent with the willingness to pay (i.e. somewhere around Rs 40–80 per household per month, the amount needed to make VESP projects financially sustainable). Nonetheless, usually 10 HP systems are used for grinding cereals in villages. An 8 kW load for eight hours of operation (four hours domestic and four hours commercial) is an ideal scenario for financial sustainability. Therefore, replacing a diesel-powered *atta chakki* with one powered with electricity is a clear win-win scenario for all stakeholders: consumers, who pay tariffs based on their affordability; the *atta chakki* owner, who saves about Rs 2 to Rs 4 per hour of operation; and the VEC, which can generate sufficient income through sale of electricity to domestic and commercial consumers to meet the expenses for sustainable operation

One example helps to illustrate the monetary benefits of the VESP in communities where it operated. Bhalupani is a small tribal village in Mayurbhanj district of Orissa where Sambandh, an NGO, implemented a VESP project among 44 households. Income generation activities promoted under a watershed project were integrated with the VESP. Accordingly, eight *donapatta* machines (which make leaf cups) having a total load of 2 kW were provided in connection with the gasifier power plant. In addition, a honey-processing unit (6 kW) powered by the gasifier plant, managed by Women's Federation of the Balupani *gram panchayat*, was set up in the village, with support from Integrated Tribal Development Agency, Government of Orissa.

The *donapatta* systems are operated by self-help groups, and promoted by Sambandh. These commercial units produce cup-plates from *saal* leaves and operate for about four hours daily in the evening for ten to 15 days a month. The Women's Federation pays Rs 60 per day for the electricity tariff when the *donapatta* unit is operational. The connection of this *donapatta* unit with the gasifier system ensures the financial sustainability of the VESP subproject, providing a net income of about Rs 57 per month. Without the revenue from this unit, the VEC would incur a loss of about Rs 500 every month. In addition, the Women's Federation makes payments of Rs 300 per month for about four to five months a year for using electricity from the power plant for the honey-processing unit, which contributes to the financial viability of the project.

Challenges

While we observed positive developments in some of the VESP projects visited, at the same time many others, if not most, experienced debilitating challenges.

Institutional Performance

Many projects faced sustainability challenges because of the dispersed nature of electricity demand in the villages; low economic activity (implying lower electricity demand); lower ability to pay of consumers (in the absence of cash income opportunities); difficulty in operation and maintenance; the VECs having limited technical knowledge; and, most importantly, weak fuel supply chains. Any combination of the above factors led to very low CUFs of between seven to ten percent in some villages, corresponding to a higher unit cost of energy production in those locations. Further, a lack of clarity about the roles and responsibilities among the different stakeholders (PIAs, state renewable energy agencies, and VECs) contributed to delays. The absence of group activities in many villages created a regulatory vacuum, especially where the VECs seldom met to review projects.

In many areas, the VECs were not adequately trained and empowered to manage decentralized projects effectively, especially where state forestry departments executed projects. The assumption in these communities was that VESP projects would be similar to any other government-supported projects, meaning that the government would take charge of the implementation process. While the VESP guidelines require at least 50 percent female members at the VEC, we found that women's participation was almost absent or minimal in most projects. Additionally, inordinate delays in installation and commissioning of the electricity generating systems severely impacted the mobilization of beneficiary communities. In Assam and West Bengal, for example, the majority of test projects overseen by the state forestry departments between 2005 and 2006 took more than four years to commission when they were supposed to take six to eight months.

Technological Performance

Performance analysis of some of the better-run projects such as in Karudoba (West Bengal), Bhalupani (Orissa), and Jambupani (Madhya Pradesh) indicates substantial variation in operational days and electricity generation over the course of several months. While there may be non-technical reasons for this variation such as low load or fuel supply issues, technological issues contributed as well. Indeed, inadequate management was found to be one of the most critical determinants of poor project performance. However, technical reasons for system non-operation have more to do with poor technical knowledge by the operators than with the technology per se. Still, the implication is that problems of operational knowledge need to be overcome to increase the operational efficiency of the VESP projects at the grass root level.

Inadequate post-installation maintenance networks (for example, the limited number of spare parts suppliers) contributed further to long lead times for fault rectification. Inadequate rural maintenance networks also tended to increase after-sales service costs and thereby threaten operational viability. Because of the remoteness of many projects, many found it difficult not only to attract suppliers but also to establish reliable after-sales service locally.

During field visits and interactions with PIAs and the community, the team also found great demand for irrigation pump sets (~ 5 to 10 HP Capacity), indicating that village demands could not be met by the installed low-capacity systems currently promoted by the VESP. On the other hand, in extremely remote areas, existing capacity was found to be underutilized. We found this underutilization largely the result of domestic load consuming only about one-third of installed capacity in the absence of any productive load. Some of the specific challenges with proper technology utilization included:

- DC motors used in the cooling and cleaning train of many of the gasifier systems not receiving required maintenance.
- Missing safety paper filters on many units. These paper filters can help lower dust intake, reduce maintenance, and increase the engine life of gasifier units.
- Malfunctioning battery charger systems in some of the subprojects. The batteries were not recharged during operation and had to be discharged after 15 to 18 hours, rendering gasifier systems non-operational. The batteries therefore had to be taken to the nearest town for recharging which took some time and contributed to substantial net downtime.
- Improperly built water tanks to supply clean water for cooling and cleaning. Most units have only one chamber in the water tank instead of two as called for in the design. Consequently, dirty water containing tar is re-circulated for cleaning gas. This results in inadequate cleaning of the gas, higher engine maintenance, and increased difficulty in operating the system.
- Unfiltered operation. In many SVO systems, oil is directly fed to the engine

without passing through a filter. This can lead to long-term engine damage and reduce the operational life of the system.

Plantations and Fuel Supply Management

Lack of adequate supply of biomass was another key constraint; not necessarily because it was not available but because of unorganized supply. In this regard, fuel supply chains for both biomass and oilseeds were found to be erratic in majority of the villages as villagers often did not receive monetary benefits for contributing a share of the total biomass, resulting in a "severe disinterest" in the VESP after the initial months. In the case of SVO based projects specifically, the oilseeds collected from forests yielded better prices in the local market than those sold for electricity generation. As such, villagers were more inclined to sell the oil seeds for cash income instead of contributing to energy production.

Financial Performance

In most of the test projects, revenue management was virtually absent. Because systems were down for most of the time, communities were reluctant to pay for service. Poor revenue flow diminished interest in maintaining system operations, creating a vicious cycle. Operator costs (varying between Rs 500 to Rs 1500 per month) also became significant since low lighting loads and the absence of productive loads diminished demand. Normal maintenance costs added to total expenditures, often overwhelming revenues that could be generated through user payments. Because the VECs did not levy any penalties for non-payments, they had no reason to keep any revenue-related records tracking project income and expenses.

Convergence with Other Development Programs

Inadequate capacity building and support to the VECs and system operators is yet another key factor negatively affecting the project. While the operators of the energy production system in the test projects seemed to be trained on operation of the systems, they were not very competent in dealing with maintenance aspects especially of the engine system.

For instance, in a majority of VESP subprojects, ICSs and biogas plants were either non-functional or were not given adequate importance, even though these technologies were critical to the VESP concept. Communities were not inclined to use biogas and ICSs because they were used to cooking on wood and were reluctant to change their cooking practices. For most villages, dung availability was low because cattle were not stall-fed in most of the projects, and, moreover, plenty of fuelwood was available at zero cash outlay. These obstacles ultimately meant that the VESP was not harmonized with other prevailing state and national development concerns and, thus, users and village leaders focused on other tasks and programs.

Conclusion

Planners conceived of the VESP as a holistic way of energy delivery based on the premise that biomass based technologies are reliable, operations can be viable, and local communities can plan and manage the projects with support from PIAs. The main strength of the VESP lies in its goal of "providing total energy security to achieve sustainable development" to remote areas through utilization of locally available renewable bio-energy resources and the direct, democratic empowerment of local communities. Locally sourced bio-energy was to provide a closed loop for energy generation, producing energy for a community in a self-sufficient way. This was intended to create financial opportunities, increase economic development, and provide a sense of ownership to local communities.

Despite its approach, VESP did not achieve its desired goals due to a mesh of discrete failures and unintentional weaknesses. Some of these setbacks include competition with other state-sponsored rural electrification projects, lack of clarity between various stakeholders, and after-sales service issues. The VESP struggled to maintain a steady supply of fuel due to a lack of organized collection, and because stronger financial incentives were in favor of selling fuel for other uses besides the generation of energy services and electricity. Most biogas technology suppliers showed reluctance to develop the post installation service network because of a low volume of activity, despite the fact that they had an interest in participating in programs such as the VESP. Community leaders seldom maintained systems properly, and revenue collection was rare. As a result of these interconnected factors, the VESP ultimately succumbed to its weaknesses.

Chapter 10

The Solar Home Systems Project in Indonesia

Introduction

The World Bank initiated their Indonesia Solar Home System (SHS) Project, which ran from 1997 to 2003, to promote the diffusion of solar PV technology in Indonesia through a market-based approach. Initially estimated to cost $118.1 million equivalent,[1] the project aimed to rapidly develop the solar PV market in the country, by reaching roughly one million rural Indonesians living in remote and isolated locations, primarily through the sales and installation of 200,000 SHS units in the provinces of West Java, Lampung, and South Sulawesi.[2] Unfortunately, it came into effect only months after the infamous 1997 Asian Financial Crisis swept into the region. Despite major revisions made to its design, the project never managed to regain momentum, and by project closing in 2003, less than five percent of the original sales target, or only 8,054 SHS units, had been installed, reaching a mere 35,000 villagers.[3]

While documentation from the World Bank and others were quick to highlight the financial crisis as the main reason for the project's shortcomings, a closer examination reveals several other circumstances, which are equally as pertinent to investigate. In laying out these surreptitious factors, this chapter provides a short overview of Indonesia's energy landscape, followed by a summary of the Indonesia SHS Project. After an explanation of some of the benefits the project was able to achieve, it then highlights the key challenges that the project faced concurrent to the financial crisis.

Description and Background

Indonesia is a vast, sprawling archipelago of more than 13,600 islands covering an area of roughly two million square kilometers or a little less than three times the size of Texas (the second largest state in the United States). Known as the

---

1 The World Bank 2001. Solar Home Systems. *Implementation Completion Report.* The World Bank, East Asia and Pacific Region.
2 The World Bank 2004. Solar Home Systems. *Implementation Completion Report.* The World Bank, East Asian and Pacific Region.
3 The World Bank 2004.

"spice islands" throughout much of its history, it is the largest country in Southeast Asia both in terms of population and size, and it is blessed with an abundance of natural resources. Indonesia is an important energy player in the region and has a wealth of untapped potential for renewable energy development as seen in Table 10.1; it is, also, one of the four largest emitters of greenhouse gases when changes to land use are taken into consideration. The archipelagic nature of the country's terrain makes decentralized solar PV technology an attractive option for rural electrification, considering the increasingly high cost of serving isolated and remote islands and villages. Moreover, as one of the first SHS projects initiated by the World Bank, the Indonesia SHS Project is an important foundation of knowledge regarding their so-called market-based approach to renewable energy.

Table 10.1   Potential and installed capacity of renewable energy in Indonesia

Type of Energy	Potential (MW)	Installed Capacity (MW)	Utilization Ratio
Large Hydro	75,674	3,854	5.0
Small Hydro	459	54	11.76
Geothermal	19,658	589.50	3
Biomass	49,807	177.80	0.36
Solar	4–6.5kWh/m2/day	5	n/a
Wind	3–6 m/sec	0.5	n/a

In 1995, the country was still riding a wave of high economic growth resulting from the dramatic increase in oil export revenues in the 1970s.[4] Moreover, the abundant oil and gas sectors were supplying over 85 percent of the country's commercial net energy consumption.[5] However, a GDP per capita of $1,014 placed Indonesia sixth out of 10 countries in Southeast Asia.[6] Approximately 17.6 percent of its 199 million people (roughly 35 million) lived below the national poverty line[7] and more alarmingly, 60 percent of all Indonesians still had no access to basic

---

4   The World Bank 2011. Interregional Resource Transfer and Economic Growth in Indonesia, Volume 1. *Poverty and Inequality.* Available at: http://econ.worldbank.org/external/default/main?pagePK=64165259&theSitePK=477894&piPK=64165421&menuPK=64166093&entityID=000009265_3980429111107 [accessed: August 15, 2011].

5   The World Bank 1996. Republic of Indonesia Solar Home Systems Project. *Project Document.* Washington: The World Bank.

6   The World Bank 2011. *Data.* Available at: http://data.worldbank.org/indicator/NY.GDP.PCAP.CD?page=3 [accessed: July 2, 2011].

7   The World Bank 2011.

electricity services.[8] The 1993 Outlines of State Policy highlighted the importance of an adequate, reliable, and reasonably priced electricity supply to serve the country's productive sectors, improve the living standards of Indonesians, and ultimately sustain Indonesia's economic and social development.[9] Thus, with 70 percent of the population still living in rural areas, expanding rural electrification was integral to the government's economic development strategy.

Throughout the 1980s and the 1990s, the power sector in Indonesia experienced rapid expansion, particularly in the main islands of Java and Bali. The State Electricity Corporation (PLN) increased their installed capacity five-fold, from 3,032 megawatts (MW) in 1981 to over 15,000 MW by 1995.[10] The company was connecting more than 1.5 million new customers a year and carried out an investment program of about $3.5 billion annually.[11] Through grid expansion, and where necessary, the deployment of isolated diesel generators, electricity access was reaching 39,000 villages, a ten-fold increase from 3,400 villages in 1980.[12]

Despite all these achievements, however, rural electrification coverage in Indonesia was still at 40 percent in 1996—well below the regional average.[13] As an illustration, neighboring Thailand and Malaysia were reporting rural electrification coverage of 80 and 98 percent respectively.[14] Grid expansion was particularly challenging outside of Java and Bali where tens of thousands of villages and hamlets known to exist at the time were sparsely scattered across thousands of islands, crisscrossing 5,100 kilometers from East to West and 1,800 kilometers from North to South.[15] Full grid-based electrification was estimated to cost as much as $5 to 6 billion per year[16]—a financial commitment that the government was not prepared to make; and in any case would take as long as 30 years to complete.[17] Nonetheless, the political and socio-economic implications of depriving 115 million Indonesians of the most basic electricity services at the dawn of a new century could not be easily ignored.

Owing to the abundance of sunlight in most parts of the country,[18] solar PV technology, particularly in its application as a SHS, had long been recognized as a viable alternative to conventional grid electricity, especially in areas where households were dispersed and energy demand was still quite low. Following the positive outcomes of several demonstration projects—including those in the villages

---

8   The World Bank 1996. Republic of Indonesia Solar Home Systems Project. *Project Document*. Washington: The World Bank.
9   The World Bank 1996.
10   The World Bank 1996.
11   The World Bank 1996.
12   The World Bank 1996.
13   The World Bank 1996.
14   The World Bank 1996.
15   The World Bank 1996.
16   The World Bank 1996.
17   The World Bank 1996.
18   An average irradiation of 4.3 kWh/m2 (Prastawa 2000).

Figure 10.1   SHS units from a government-funded program on Lake Cirata, West Java, Indonesia

of Sukatani and Cileles in West Java—the Indonesian government initiated the Solar Power for Rural Electrification scheme (*Listrik Tenaga Surya Masuk Desa*) in 1991, in which 3,545 SHS units were successfully deployed in 13 provinces. By the mid-1990s, approximately 20,000 SHS units had been installed throughout the country, mainly through government-funded programs backed by international development donors. An evaluation of these efforts indicated that users were generally satisfied with the performance of their SHSs and did not experience major problems with critical components such as batteries, panels, and controllers.[19] During our field visits in West Java and Lampung, we had the opportunity to interview some of the users who had benefited from the government largesse—many of whom had been using their SHSs for the past ten to 20 years, like the one shown in Figure 10.1.

In 1995, however, a local entrepreneur in West Java managed to sell 4,000 SHS units on credit, in the first year of operation,[20] despite such ongoing government-funded programs for SHSs. This encouraging development was consistent with the

---

19   The World Bank 1996.

20   Miller, D. and Hope, C. 2000. Learning to Lend for Off-grid Solar Power: Policy Lessons from World Bank Loans to India, Indonesia, and Sri Lanka. *Energy Policy* 28 (2000), Elsevier Science Ltd.

success of pioneering SHS companies in rural Kenya in the early 1990s as well as experiences in the Dominican Republic, Sri Lanka, and Zimbabwe.[21] Seemingly, technological innovations coupled with the availability of compatible and energy-efficient devices had made the SHS market more competitive. Thus, in the absence of grid connection, the lesson appeared to be that rural households were willing to pay market prices for a reliable alternative.

These market trends convinced planners to embark on "something big" to simultaneously validate the World Bank's energy strategy and meet Indonesian rural energy targets. The Indonesia SHS Project came about in 1996 as part of a larger endeavor by the World Bank to promote the commercial diffusion of SHSs as a cost-effective alternative to grid expansion in developing countries. Specifically, it would be feeding into the implementation of the government's "50 MWp One Million Roof Program"—an initiative to install one million SHS in rural households by 2005.[22] Although the proposal for a new SHS project hinged on the credibility of the World Bank as the largest financial lender in the power sector, the World Bank's experience had actually been predominantly one of lending for large centralized plants or grid extension projects. In fact, the only relevant experience that it had at the time was the ongoing India Renewable Resources Development Project launched in 1994, and it was already experiencing major difficulties including the risk-averseness of lending banks in financing rural credit, a lack of a market infrastructure, and inadequate support for the private sector.[23] The Indonesia SHS Project nevertheless set an ambitious target of selling and installing 200,000 SHS (10 MWp) to supply electricity to approximately one million rural villagers. Table 10.2 documents the main historical milestones of SHS deployment in Indonesia prior or concurrent to the launch of the project.

The Indonesia SHS Project ran from 1997 to 2003 and was valued at $118.1 million equivalent, with seed money of $44.3 million equivalent or 38 percent of the project costs to be provided by the World Bank and the GEF.[24] It was to be a massive undertaking, requiring serious investments to be made into both developing Indonesia's solar PV market and formulating a national energy access policy to incentivize the adoption of renewable energy technologies. However, rather than relying on government funding, the bulk of the project's costs of $67.3 million was to be financed mainly on credit from sub-borrowers (SHS dealers) and end-users (rural customers) as summarized in Table 10.3. The idea was to

---

21   Miller, D. and Hope, C. 2000.

22   Retnanestri, M. et. al. 2003. Off-grid Photovoltaic Application in Indonesia: A Framework for Analysis. *Destination Renewables*. Sydney: University of New South Wales.

23   Although the Project Performance Assessment Report (PPAR) rated the overall project outcome as "satisfactory", the program did not fully succeed in reaching the rural market or in developing marketing and financing mechanisms based on cost recovery principles.

24   The World Bank 2004.

Table 10.2   SHS deployment in Indonesia, 1988–1997

Year	Name of initiative	Location	Sources of funding	(Targeted) SHS units deployed
1988–1989	Sukatani Solar Project	Sukatani, West Java	GOI and R&S Eindhoven	102
1988–1992	Solar Power for Rural Electrification Scheme (Listrik Tenaga Surya Masuk Desa)	Thirteen provinces.	Presidential Aid Program (BANPRES)	3,545
1997–2005	50 MWp One Million Roof Program	Multiple provinces.	Multiple sources.	1,000,000
	– AUSAID Project (1997–1999)	Nine provinces in Eastern Indonesia.	GOI and AusAID	36,400
	– e7 Project	n/a	GOI and e7	1000
	– Indonesia SHS Project (1997–2002)	Lampung, West Java and South Sulawesi	GOI and World Bank/GEF	200,000
	– French Government Project	n/a	GOI and France	1,300
	– Bavarian-Indonesian Government Solar Project (1997)	East Java	GOI and Bavarian government	35,000 (and 300 solar village centers)

target only those villagers willing and able to pay for electricity services in order to nurture and develop a self-sustaining solar PV sector.

Credit Component

The main part of the project was the credit component, which sought to extend electricity services to about one million people through the sale and installation of 200,000 50 Watts-peak (Wp) SHS units to rural households and small commercial establishments such as the one depicted in Figure 10.2. A $20 million equivalent International Bank for Reconstruction and Development (IBRD) loan channeled through four commercial participating banks (PBs) provided a credit facility to address the high cost of SHS units and the financial constraints of dealers and potential customers.

Rural areas that could not expect grid connection from PLN in the next three years or more were identified in the provinces of West Java, Lampung, and South Sulawesi as potential regional markets, with the intent of including North Sumatera at a later stage. All these provinces had rural communities with strong purchasing

Table 10.3   Sharing of project costs

Stakeholder	Project cost ($m)	% of Total
World Bank (IBRD Loan)	20.0	17
GEF Grant	24.3	21
GOI	1.5	1
PB	5.0	4
Sub-borrowers/End-users	67.3	57
Total	118.1	100

power due to cash crops such as coffee, cacao, and palm oil.[25] West Java was additionally selected due to the initial success of the local entrepreneur mentioned above and also because of proximity to the capital, Jakarta. A population of 38 million easily made it the most populous province in Indonesia at the time, with 19 million people still waiting for electricity and other critical infrastructure.[26]

As PBs lacked the rural networks to deal directly with customers, a dealer-sales model was employed, whereby six Jakarta-based dealers were tasked to establish rural outlets and would take responsibility for the procurement, sales,

Figure 10.2   A small commercial establishment powered by lights from an SHS, Indonesia

---

25   The World Bank 2004.
26   Retnanestri, M. 2007. *The I3A Framework: Enhancing the Sustainability of Off-Grid Photovoltaic Energy Service Delivery in Indonesia.* Sydney: University of New South Wales.

installation and maintenance of SHS units; and for offering term credit to make the systems more affordable to prospective customers. The eligibility criteria for dealers included proven business competence, the existence of sales or services infrastructure in the targeted markets, and a credit agreement with a PB.[27]

A project-approved 50 Wp SHS unit with the necessary components, the only one eligible under the program, cost between $550 and $800, depending on the sales location.[28] Dealers would typically offer credit to prospective customers based on a first cost buy-down in the range of $75 to $100, funded by a separate GEF grant mentioned below. This would bring down the unit cost balance to a level that could be paid in monthly installments over a period of four to five years, in amounts roughly comparable to conventional monthly energy expenditures for kerosene. Customers would in turn be responsible for servicing their own systems, although dealers could provide service contracts or guarantees for a limited period.

The World Bank estimated that credit installments and the interest generated would provide approximately $66.8 million equivalent of the project costs. In addition, a GEF grant of $20 million equivalent, translating into a first-cost subsidy ranging from $75 to $125 for every SHS unit sold, would be awarded to the dealer upon extending credit to customers. This benefit could either be passed on to customers to make the SHS units even more affordable or be used to further develop the business (for example to recruit new staff, establish new rural outlets, or expand product inventory).

Technical Assistance Component

Approximately $4.1 million equivalent was dedicated toward establishing a Project Support Group (PSG) under the authority of the government's Agency for the Assessment and Application of Technology (BPPT). Although the PSG did not directly manage the project's financing, it functioned as the coordinating body for most project activities as well as the main interface for the stakeholders summarized in Table 10.4. It worked with the BPPT to handle the recruitment and selection process for dealers; verify dealers' compliance regarding installment of equipment; monitor proper utilization of the GEF grant; provide information regarding technical and financial benefits of SHS as well as risks; protect prospective and actual customers; and conduct training for stakeholders in the form of conferences, workshops, seminars, and study tours.

---

27   Martinot, E., Cabraal A., and Mathur, S. 2001. World Bank/GEF Solar Home System Projects: Experiences and Lessons Learned 1993–2000. *Renewable and Sustainable Energy Reviews 5* (2001), Elsevier Science Ltd.

28   The World Bank 2004.

Table 10.4   Interface between PSG and other stakeholders

Functions of PSG/ stakeholders	Donors (World Bank/ GEF)	Government (MEMR, BPPT, BAPPENAS)	Private sector (PBs, dealers)	End-users (customers)
Coordination	– Liaise between the World Bank and the government agencies. – Provide project reporting.	– Work with BPPT to coordinate project implementation and interactions with all stakeholders. – Undertake project level reporting.	– Recruit and select dealers.	– Provide a communication platform between actual customers and prospective customers.
Capacity Building		Limited training for selected staff from BPPT, ministries and other relevant government agencies.	– Limited training for selected staff of SHS dealers. – Provide services for business development.	– Provide information and technical assistance regarding SHS. – Provide information regarding relevant government policies (e.g. the future availability of PLN services).
Monitoring and Evaluation		– Commission studies and assessments to monitor and evaluate progress of project.	– Verify compliance of dealers. – Record financial transactions. – Monitor proper utilization of GEF grant. – Channel feedback from PBs and dealers.	– Channel feedback from customers.

In view of the project's longer-term objectives "to strengthen Indonesia's institutional capacity to support and sustain decentralized rural electrification using solar photovoltaics,"[29] about $1.2 million equivalent was allocated for policy support and about $1 million equivalent was allocated for institutional development. This involved providing assistance to the government's Rural Electrification Steering Committee to develop the "Decentralized Rural Electrification Study and SHS Implementation Plan".[30] The funds would also

---

29   The World Bank 1996.

30   A strategy and a 10 year-action plan to meet the modern energy needs of the rural population in Indonesia through renewables, where solar PV technology represents a least

be used to strengthen the capacity of the BPPT and the Ministry of Energy and Mineral Resources to develop technical specifications, and to carry out type and product testing, certification, and monitoring of SHS units. Table 10.5 summarizes the allocation of project funds to the different project components and the barriers that it was expected would be overcome.

Table 10.5   Project components addressing different barriers

Description	Project cost $m	% of total	Demand side barrier(s) to be addressed	Supply side barrier(s) to be addressed
Credit	111.8	95	High transaction costs. Lack of credit facilities.	Lack of dealers and strong supply chains.
Implementation Support	4.1	3	Lack of information regarding benefits and risk of the technology. Unfamiliarity with the type of investment/financial model.	Lack of in-country experience in organization and financing.
Policy Support	1.2	1		Lack of policy framework to support penetration of solar PV technology in the long term.
Institutional Development	1.0	1		Lack of institutional and capacity to disseminate solar PV technology in both the short- and long-term.

Benefits

Despite its inability to accomplish its goals, the project has yielded some benefits such as raised awareness regarding solar PV technology; delivery of a minor but measurable amount of clean, modern, and affordable electricity services; and improvements in the capacity of some stakeholders.

Raised Awareness

Although rural Indonesians had been exposed to solar PV technology through government-funded SHS programs as early as the late 1980s, the Indonesia SHS Project introduced the concept of commercial value. Due to limited funds, dealers were not able to afford TV or radio commercials or even brochures. Thus, in order to reach as many people as possible, usually a technician would

cost option.

make a presentation in each village community center, followed by a technical demonstration. "It was always a very formal affair," explained one respondent. "It was very important to ensure that the village chief is present in this presentation, to give him respect. If you are able to convince him regarding the importance of the SHS and the legitimacy of your business, it is easier to approach and educate other villagers." These marketing campaigns, scarce and homegrown as they were, were not only opportunities for villagers lacking electricity services to the learn more about solar PV technology and SHSs; they also empowered them to assess and prioritize their energy needs and decide on an option for a reliable, autonomous, and environmentally-friendly source of electricity.

At the institutional and policy level, the project served as a reference point for policymakers working on rural electrification projects involving solar PV technology. From 2008 to 2010, for example, the government spent an average of $100 million per year to further diffuse SHS in the country and attract private sector participation through programs in various ministries including the MEMR, the KKP, the KPDT and provincial and local governments.[31]

Delivery of Energy Services

The project appraisal in 1996 estimated that as many as 62,000 households –39,000 located outside Java and Bali—did not have access to basic electricity services within the project's geographic service area.[32] After a slow start, the gradual reduction of kerosene subsidies from 2000 as seen in Table 10.6 increased the competitiveness of SHSs leading to a significant spike in sales. This was followed by an increase in value of the main cash crops relative to the price of one SHS unit starting in 2001.[33] By the close of the project in 2003, approximately 8,500 households, or about 30,000 customers were benefiting from the delivery of electricity services provided through SHSs.

Though the SHS program was limited in the number of villages it reached, those it did were pleased with their systems. The villagers we interviewed during our field visits in West Java and Lampung all confirmed their satisfaction. Among the most cited benefits are the relative affordability of SHSs compared to having to pay for monthly purchases of kerosene or diesel; the ease in which the systems can be maintained and operated; and the entertainment and communication value derived from being able to use radios, TVs, and mobile phones. During our field visit to Lake Cirata, we were also able to observe the usage of SHSs for income-generating activities in the fish-farming industry and other small commercial establishments.

---

31   Respati, J. 2010. The Dilemma of Solar PV Utilization in Indonesia. *Respect.* Jakarta: Respect.

32   The World Bank 1996.

33   Except for coffee prices which actually plunged to their lowest since 1973 (Retnarestri 2003).

Table 10.6   Overall SHS sales in comparison with major cash crop sales

Description/Year	1997	1998	1999	2000	2001	2002	2003
Exchange rate ($1 to IDR)	3116	9501	7782	8470	10,411	9,549	8,577
SHS Cost (IDR)	1 m	3 m	3.1 m	3.2 m	3.3 m	3.4 m	3.5 m
SHS Sales (unit)	0	0	92	1299	1552	972	4139
Kerosene price per litre (IDR)	250	250	250	350	400	600	900
Palm oil (Kgs/1 SHS)	8,930	10,158	8,497	13,770	11,443	9,127	8,122
Coffee (Kgs/1 SHS)	423	220	393	1,018	1,160	1,262	901
Cacao (Kg/1 SHS)	493	281	392	545	499	462	357

Stakeholder Capacity

During the course of the project, BPPT, as its primary implementer, was able to expand their know-how of solar technology and become the focal point for solar technology development in the country. Their achievement in developing strict technical criteria and procedures to test and certify SHS units has been adapted in other developing countries such as Sri Lanka, China, and Uganda.[34] In addition, BPPT's PV laboratory successfully obtained ISO 17025 accreditation for testing and certifying balance of system components.[35] Junior engineers, in particular, benefited immensely from their training and immersion in this laboratory. The PSG, contracted by the BPPT to manage project activities, was also able to build capacity in technical assistance and project monitoring and evaluation.

Although Indonesia's solar PV industry remains relatively underdeveloped in comparison to other developing countries, the project did manage to include 479 technicians working for SHS dealers in market development provided by the PSG, and in coaching and business implementation frameworks provided by the World Bank. The project also successfully established a market supply chain of more than 100 dealer outlets by 2003.[36] There was a reactivation of the Association of Indonesian SHS Dealers in 2000, which worked on establishing an accreditation system and setting minimum quality standards for SHS dealers.[37] Toward the end of the project, dealers were assisted in contacting potential investors and

---

34   The World Bank 1996.

35   The World Bank 1996.

36   Retnanestri, M. et. al. 2003. Off-grid Photovoltaic Application in Indonesia: A Framework for Analysis. *Destination Renewables.* Sydney: University of New South Wales.

37   GEF 2004. Indonesia Solar Home Systems. *Terminal Evaluation Review Form.* Global Environment Facility.

funding sources, including the Solar Development Fund, which was at that time in discussions to develop a partnership with Bank Rakyat Indonesia (BRI) to assist SHS dealers.[38] Essentially, project documents cite an "enabling policy environment" and "business enterprise support" as some of its main positive outcomes.[39]

Challenges

As surmised above, the Indonesia SHS Project seemed ready to promote solar PV and SHSs through the implementation of its credit and implementation support components, as well as through its assistance to the solar PV industry. However, soon after the project became effective in October 1997, it became clear that its design needed a major overhaul owing to the rapidly deteriorating economic and political situation in Indonesia following the Asian Financial Crisis. The devaluation of the Indonesian Rupiah (IDR) against the US Dollar had resulted in a severe credit crunch in the banking sector, "the worst since the 1970s," according to one respondent. Two of the four PBs closed down; whereas the other two were barred by Bank Indonesia from offering credit until 2000.[40] Concomitantly, the high import content of SHS units had increased their price more than three-fold, hampering the ability of both dealers and potential customers to sell or buy SHS units.

   Starting from 1998, significant changes were made including revising sales targets from 200,000 to 70,000 SHS units; reducing the standard size of the SHS units sold from 50 Wp to a minimum of 10 Wp; adjusting the GEF grant to a \$2 per Wp subsidy instead of a per system subsidy; closing the IBRD loan due to lack of demand for credit; and canceling the "Decentralized Rural Electrification Study and SHS Implementation Plan".[41] Unfortunately, these measures proved ineffective. The World Bank admitted that the project became plagued by "slow progress of the SHS sales, weak investment in rural distribution networks, and inability of the banks to make loans to SHS dealers"[42]. Interestingly, these challenges mirrored almost exactly the difficulties that the World Bank was already facing in its ongoing 1994 India Renewable Resources Development Project. Our interviews with key stakeholders reflecting on the project almost fifteen years onwards suggest that the reasons for project failure may have been more fundamental, and that perhaps the financial crisis became an excuse rather than a major impetus.

   We found that the project's credit component was ill equipped from the beginning to help the fledging solar PV industry overcome first cost hurdles. This was mainly due to a poorly conceived credit facility that failed to provide the suitable financial infrastructure and banking products for a rural clientele, and to

---

38   GEF 2004.
39   The World Bank 2004.
40   The World Bank 2004.
41   The World Bank 2004.
42   The World Bank 2004.

support struggling SHS dealers. Despite claims of sustainability, it is clear that the project remained oriented to the short-term, with little scope to truly contribute toward the development of Indonesia's PV industry.

Poorly Designed Credit Vehicle

At the start of the project, Indonesia's solar PV market was what the World Bank characterized as in a "high price low volume" equilibrium.[43] As SHSs are self-contained generation and distribution systems, the initial capital cost is very high in proportion to the total operating and maintenance costs over the lifecycle—in many cases, representing almost one year of income in low- and middle-income rural households.[44] Moreover, under current Indonesian banking practices, commercial banks were only allowed to offer credit over a period of one or two years, which is hardly an affordable cost amortization period for such households. Despite the various measures that had been put into place under the project, "a lack of established high-volume supplier-dealer chains, high prices, and a lack of term credit" continued to hamper market development.[45]

As mentioned above, a $20 million equivalent IBRD loan was channeled through Bank Indonesia to four PBs to provide SHS dealers with access to capital investment and to allow them to offer credit lines to prospective customers. Due to repercussions of the Asian Financial Crisis, however, two of the selected PBs were not able to participate due to their dire financial situation; whereas the other two remained wary of Bank Indonesia's increasingly strict regulations on non-performing loans (NPLs) even after their recapitalization was completed in the middle of 2000.[46] In the end, only one PB was prepared to offer any credit, and despite keen interest from SHS dealers, only $0.1 million of the $20 million loan was utilized before the World Bank decided to close it down at the end of 2000, fifteen months ahead of schedule.[47] Subsequently, five out of the six dealers that had committed to the project went out of business.

The design of the credit facility focused too much on mobilizing SHS dealers and too little on aligning to the priorities and concerns of PBs and building their capacity as the managers of the funds. Apart from the financial crisis, the risk-averseness of the PBs was also due to their lack of familiarity with the rural market in general and solar PV technology in particular. Serving rural customers with limited income and assets would have required experience in rural banking products such as microfinance, as well as a strong presence on the ground, the

---

43   The World Bank 2004.

44   Cabraal, A. et. al. 1997. *Accelerating Sustainable PV Market Development*, The World Bank.

45   The World Bank 2004. Solar Home Systems. *Implementation Completion Report*, The World Bank, East Asian and Pacific Region.

46   The World Bank 2004.

47   The World Bank 2004.

Figure 10.3   A small cooperative on Lake Cirata, West Java, Indonesia

collective domain of the thousands of government cooperatives and microfinance institutions, one of which is pictured in Figure 10.3. In addition, PBs would have to experiment with a business model they did not understand. "Renewable energy projects are very risky compared to coal projects," claimed one respondent, "We do not have the requisite knowledge to finance them." Another admitted, "We would not know what to do with reacquired SHSs in the case of defaulting customers, unlike with motorcycles," referring to the popularity of credit lines for motorcycles.

At the same time, it appears that potential customers, at least in some target communities, did not properly understand the supposed benefits of the credit facility. Among the SHS users we interviewed, some had made use of the available credit to pay for their systems, but an equal number of respondents had paid cash, as they were unfamiliar with the banking practices in general. These respondents generally represented households that were in the upper-income bracket of the rural population. With more disposable income, they typically had larger SHS units and used the electricity for some productive uses such as lighting fishponds or small convenience shops. They were also often former owners of diesel-powered generators, glad to be using more economical systems.

However, we also encountered those respondents from lower-income households that had little or no source of lighting prior to their SHS units and had benefitted from either free government-funded SHS programs and/or the cheaper second hand SHS market rather than from participating directly in the project. One of these respondents commented that had it not been for the free SHS, he would have not minded to continue living in darkness. When combined with what PBs viewed as excessive bureaucratic borrowing requirements imposed by the World Bank, it is understandable why PBs considered the project "doable, but not bankable" and therefore impractical to warrant robust involvement.

Instead, the project placed the burden of SHS commercialization almost entirely on inexperienced SHS dealers through its dealer-sales model. The project's credit component as it stood made dealers and customers responsible for financing $66.8 million equivalent or almost 60 percent of project costs, mainly through the payment of monthly installments. It seemingly distributed the investment risks of the credit facility among the different stakeholders involved, with the PBs bearing the dealer credit risk and the dealer bearing the consumer credit risk. However, because it was the dealers rather than the PBs, the World Bank, or the government that were responsible for the complex and arduous task of administering the loans to customers and monitoring compliance, it was also ultimately the dealers who were left shouldering the financial risk of loan defaults. "It would have been far preferable for us that banks be in charge of the loans," mentioned one respondent. "When banks are responsible for collecting payments, companies can focus on providing the SHS and related services."

Inadequate Dealer Support

SHS dealers were mainly small and inexperienced enterprises in a nascent market, peddling an unfamiliar product and novel concept of electricity services. Deprived of their main source of investment capital from the very beginning due to the reluctance of PBs to offer credit, dealers were further constrained in their ability to finance and develop their businesses as the price of SHSs jumped three-fold following the drastic depreciation of Indonesian currency. This was especially true after the IBRD loan was terminated and dealers only had the option of using their own financing to continue their businesses. "Without credit from the banks, we had to provide financing from our own pockets," explained one respondent "This was very tough for small businesses like ours." Even when sales targets were reduced from 200,000 to 70,000 units in 2001, dealers were still not able maintain sufficient inventories and establish the necessary rural outlets. As described by another respondent, "I had to cover three whole regencies with only one motorbike. It was an impossible job."

However, rather than being allowed to focus on building a proper SHS supply chain and a rural service infrastructure, dealers also had to build their rural credit delivery and collection infrastructure—both requiring very different sets of skills and expertise. In this context, respondents felt strongly that "the magnitude of the installation targets was not comparable with the efforts to build capacity." Apart from a few workshops that were limited to sending only a few staff at a time, there was very little support for dealers to upgrade their skills and expertise, develop their businesses, approach banks for financing, learn about rural credit, and address problems on the ground. The grants provided by GEF did little to improve their "unsatisfactory" performance as the project required that dealers offer credit to their customers as a condition of eligibility to receive these grants. This caused problems for dealers who did not feel secure enough to borrow or extend credit. Moreover, as a respondent lamented, "A $100 for every SHS sold is not enough. They should have increased the grant amount after the crisis."

High Prices

Lacking a workable credit facility to make systems more affordable and a proper supply chain to reduce transaction costs, the development of the solar PV market was further impeded by several factors that the project design was not able to rectify. Certainly, the financial crisis affected the purchasing power of many potential customers that saw slumps in the value of their cash crops. However, most respondents we interviewed criticized the continued prevalence of free SHSs provided through government-funded programs in parallel to the project. Some villagers we talked to in Lampung mentioned that they had preferred to wait for these free SHSs, even though stocks were limited and the waiting lists were long, rather than purchase their own units. Other villagers had continued to use kerosene lamps, benefiting from highly politicized government subsidies that were only stopped in 2000.

A lack of coordination with PLN was another problem as former customers living in target areas that were eventually abandoned by dealers due to the availability of grid electricity flooded the market with cheaper and less-regulated second-hand SHSs. Many villagers we interviewed during our field visits admitted that they had gotten their systems from this second-hand market. These respondents stated that they preferred to receive inferior goods rather than pay the premium for a new system. Considering the well-known fact of inadequate after-sales services—which at some point became practically non-existent after all but one dealer remained in business during the project—it was perhaps not a poor choice to make.

Most damaging, however, was that the project had to bear foreign exchange costs of imported SHS components amounting to approximately $85 million equivalent, or more than 70 percent of project costs.[48] Solar panels, the most expensive component, were imported from Japan, Korea, and Germany. Some SHS parts such as charge controllers, batteries, and energy-efficient bulbs were already being produced domestically at the time, yet contained a significant amount of imported parts and materials.

Unsurprisingly, dealers used the opportunity of the financial crisis to venture into foreign solar PV markets and benefit from the much stronger US Dollar. The *Implementation Completion Report* cites the success of certain dealers in exporting balance of system components to Sri Lanka as part of the World Bank's Energy Services Delivery program (see Chapter 8!) as well as for commercial sales in

---

48   The World Bank 1996.

Kenya.[49] Respondents felt that instead of subsidizing foreign PV markets, the project could have invested some of this funding into developing the domestic solar PV assembling and manufacturing industry, which would have gradually brought down dependence on imports and consequent high SHS costs. Although the BPPT did make some inroads in this direction, the industry is still underdeveloped today, very much dependent on imported content, and so far unable to reap the benefits of economies of scale, despite the fact that the country has recovered remarkably from the financial crisis.

Insufficient Government Involvement

From the analysis above so far, it is clear that although the main objective of the project was to catalyze Indonesia's solar PV market, the private sector was not ready to take a lead role in its implementation. The four PBs that had been selected were still unfamiliar with investments in the renewable energy sector and none of the six appointed SHS dealers had developed an effective supply chain and financial mechanism to deploy SHSs on the scale intended by the project. The solar PV market was still very much in its infancy and the project therefore needed greater government involvement to guarantee appropriate the institutional and regulatory environment.

The World Bank selected the BPPT under the Ministry of Research and Technology (MENRISTEK); the Directorate-General of Electricity and Energy Development (DGEED) under the MEMR; the National Planning Agency (BAPPENAS); the Ministry of Finance (MENKEU); and the Ministry of Cooperatives and SMEs (KKP), as government stakeholders to guide the implementation of the project. In particular, the BPPT played the important role of main executing partner. However, despite what could be perceived as strong government support, the project was seriously hampered by a lack of coordinated involvement among these different agencies and their relevant counterparts. In fact, no government institution took on the role of oversight, overall coordination, and regulator. Project documents cite BPPT's performance as "satisfactory" and even "exceeding expectations,"[50] when in reality BPPT focused only on technology development and standards and did not concern itself with other aspects of the project such as profitability, investment opportunities, stakeholder coordination, or marketing and supply chain logistics.

In fact, it is rather surprising considering the lack of experience on part of both the World Bank and the Indonesian government that the project only set aside five percent of costs for technical assistance and capacity building purposes as reflected in Table 10.7. As a comparison, between 2000 to 2008, the World Bank was spending in aggregate about one-quarter of investments or $1 billion in supportive investments in energy access, much of which went to the development of public sector capability

---

49   The World Bank 2004.
50   The World Bank 2004.

such as rural electrification master plans, policy frameworks, and energy strategies.[51] The government's own in-kind commitment toward the project through BPPT only represented a total of one percent of project costs and was significantly reduced with the cancelation of the "Decentralized Rural Electrification Study and SHS Implementation Plan" (a lost opportunity to create a more lasting policy framework for the solar PV commercialization). Respondents we interviewed suggested that many of the project's shortcomings could have been addressed if there had been a more serious commitment from the government to oversee the implementation process, and from the World Bank to build institutional capacity.

Table 10.7   Allocation of funds to the different project components

Description	Project Cost $m	% of Total
Credit Component	111.8	95
World Bank (IBRD Loan)	*20.0*	*17*
GEF Grant	*20.0*	*17*
Participating Banks	*5.0*	*4*
Dealers/end-users	*66.8*	*57*
Implementation Support	4.1	3
GEF Grant	*3.1*	*3*
Government	*0.5*	*1<*
Dealers/end-users	*0.5*	*1<*
Policy Support	1.2	1
GEF Grant	*0.7*	*1*
Government	*0.5*	*1<*
Institutional Development	1.0	1
GEF Grant	*0.5*	*1<*
Government	*0.5*	*1<*

Lack of Project Sustainability

Considering the longer-term objective of the Indonesia SHS Project to advance renewable energy commercialization and create a niche market for solar PV technology, it did not provide many building blocks to sustain the market after it closed. For example, BPPT's success in testing and certification of SHSs did not translate into better capacity building opportunities for other stakeholders. "BPPT was in a very privileged position. As the focal point of the project, it benefited from all capacity building efforts. But it did not encourage other elements of the market to grow," criticized one respondent. The premature closing of the IBRD

---

51   Barnes, D., Singh, B., and Shi, X. 2010. Modernizing Energy Services for the Poor: A World Bank Investment Review – Fiscal 2000–2008. *World Bank Energy Sector Management Assistance Program.* Washington: The World Bank,

loan, which resulted in all but one dealer going out of business, also indicates that dealers were not successful in developing the capacity to enter the market without project support let alone being able to independently catalyze commercial demand for solar PV technology. "There was a large vacuum in the solar PV market until 2005," described another respondent.

In addition, respondents criticized monitoring and evaluation of project impacts as insufficient. For example, there was no study carried out to properly measure how the financial crisis really affected the credit component of the project. Very little information was provided regarding the state of SHSs installed through the program. "We estimate that most of the SHSs installed have not been in operation for a long time. However, there is no data to back this up," stated one respondent. There has also not been a study to assess whether the technology has been understood and accepted by the wider Indonesian population. In fact, many questioned the choice of SHSs in the first place, perceived by some as an unfamiliar technology "imposed on Indonesia" from the World Bank rather than a "need stemming from an expressed interest of the rural population." Doubts were raised whether the technology was even suitable for sufficient solar irradiation considering frequent cloud cover, high levels of humidity in the tropics, and the fact that many of the remote areas targeted are in dense forest areas. In this regard, it was suggested that perhaps concentrated solar power or other renewable energy sources such as geothermal, hydro, or biogas could have been more appropriate solutions for rural electrification.

In fact, some respondents question the sole emphasis on electrification, which in their opinion emphasized consumptive and leisure activities rather than productive uses of energy. It was suggested that the project would have been more impactful had it considered investing in other important rural energy needs such as mechanical power, cooking, transportation, and telecommunications that do not necessarily depend on better electricity services. As an illustration, 72 percent of the population or 156 million Indonesians currently rely on biomass for cooking and heating.[52] Investing in better cooking stoves would have had immediate and significant impacts on household welfare in terms of improving health and reducing the hours spent on firewood-related drudgery that could be better used for more productive activities.

The provinces chosen for potential target markets were also in question considering not many dealers had networks in Lampung and South Sulawesi at the beginning of the project. As a result, most were functionally excluded from taking part. Some respondents were of the opinion that the selection of the target areas was too ambitious, whereas others thought that the project could have included more provinces and did not do enough to leverage on the natural strongholds of many other competent SHS dealers.

---

52   Barnes, D., Singh, B., and Shi, X. 2010.

Conclusion

The credit component of the Indonesia SHS project was designed to overcome the first cost barrier found to be critical in the uptake of capital-intensive technologies such as SHSs in rural areas.[53] However, no stakeholder was willing to fully shoulder the investment risks of an unchartered rural market for solar PV technology. What was needed was a responsive financial infrastructure that provided several mechanisms to reduce the risk of investing in a new market and that allowed for greater flexibility to adapt business models to changing market signals. The Indonesia SHS Project did none of these things, and it reveals the necessity of coordinated government support for energy access and development programs. It would have benefited from the integration or harmonization of policies, the creation or alteration of institutions, and the development of well-targeted subsidies or tax structures that incentivize rather than hinder businesses.

The confluence of these three factors—a flawed financing vehicle, poorly structured SHS supply chain, and fragmented government policy—remind us that promoting energy access and responding to energy poverty are highly contextual. Without the proper alignment of economic, technical, and political factors, even those projects with the best intentions can quickly fail to achieve their goals.

---

53   Miller and Hope 2000.

Chapter 11

The Small Renewable Energy Power Program in Malaysia

Introduction

The Malaysian Small Renewable Energy Power (SREP) Program attempted to install 500 MW of biomass, biogas, municipal solid waste, solar PV, and mini-hydroelectric facilities from 2001 to 2005, but ended up achieving only 12 MW of capacity by the end of 2005. Malaysian planners altered the SREP by lowering its target to 350 MW and extending it for another five years, but by the end of 2010 just 11 projects and 61.7 MW of capacity had been built.

This chapter explores the history of the SREP in Malaysia, its drivers and benefits, and the challenges planners faced when implementing it. The SREP was a central component of Malaysian energy policy, and thus it provides an ideal situation to explore the dynamics at work within national energy planning. The SREP was specifically the cornerstone of the country's Fifth Fuel Diversification Plan and also featured prominently in the Eighth Malaysia Plan (2001 to 2005) and the Ninth Malaysia Plan (2006 to 2010). Investigating the SREP offers a deeper understanding of the pressures and interests related to Malaysian energy policymaking, and it reveals a thorny web of barriers involved in expanding renewable energy access.

Description and Background

Although Malaysia relies primarily on fossil fuels to meet national energy and electricity demand, the country is blessed with abundant renewable resources. One peer-reviewed study estimated no less than 30,700 MW of technical renewable energy potential shown in Table 11.1, when in 2009, *existing* installed capacity was less than 23,000 MW.[1] The IEA has also projected that realizable potential for renewables in Malaysia is about 130 Terrawatt-hours (TWh) per year by 2030, but that in 2006 the country generated only 101.3 TWh from its entire

---

1   Oh, T.H., Pang, S.Y., and Chua, S.C. 2010. Energy Policy and Alternative Energy in Malaysia: Issues and Challenges for Sustainable Growth. *Renewable and Sustainable Energy Reviews* 14, 1241–1252.

installed capacity—inclusive of fossil fuels *and* renewable sources of supply.[2] Both estimates imply that Malaysia could be completely powered by renewable resources, presuming they cover baseload electricity needs or are coupled with energy storage.

Table 11.1   Achievable renewable energy potential for Malaysia (MW)

Hydropower	22,000
Solar photovoltaics	6,500
Biomass and Biogas	1,300
Mini-hydro	500
Municipal solid waste	400
Total	30,700

Malaysian planners seem cognizant of this potential, and embarked in 2001 to capture some of it through the SREP. One expert we interviewed explained the decision to proceed with the SREP as follows:

> We had the four fuel policy operating for many years, and it was basically successful at promoting large hydro and a collection of natural gas and coal-fired power plants. But we saw the need to promote other types of renewable energy. Though we had some diesel power plants, these were inefficient, costly, and polluting. Most of the large hydro potential had either already been tapped, or was in places like Sabah and Sarawak, hundreds of kilometers away from major urban centers. Increased reliance on coal was thought to be too environmentally damaging, and was sometimes opposed by local communities. Natural gas was already an uncomfortably large share of the national electricity portfolio. Renewables were seen as the only viable alternative.

The Malaysian government also recognized that a collection of pernicious barriers prevented the wider adoption of renewable energy in the late 1990s and early 2000s. These included lack of a national policy in support of renewable energy, the perception that waste-to-energy and palm oil technologies were polluting, the inability to cover project costs and lack of financing, and poor coordination among different national players and ministries. One study went so far as to argue that renewable energy was looked upon as a "primitive and dirty fuel."[3]

---

2   Olz, S. and Beerepoot, M. 2010. *Deploying Renewables in Southeast Asia: Trends and Potentials*. Paris: International Energy Agency/OECD.

3   Koh and Kong 2002, 36.

Figure 11.1   The 2 MW Kerling minihydro plant, Malaysia

To address these issues, the Ministry of Energy, Green Technology, and Water announced the SREP on May 11, 2001. It was intended to be the main vehicle to meet the renewable energy targets espoused by the Eighth and Ninth Malaysia Plans as well as the Third Outline Perspective Plan. Eligible technologies for the SREP were limited to biomass, biogas, municipal solid waste, solar PVs, and mini-hydroelectric facilities such as the one shown in Figure 11.1. For some reason, wind energy was excluded from the program, though we were never given a reason why.

The Ministry of Energy established a Special Committee on Renewable Energy (SCORE) to oversee the program and defined eligible projects as those up to 10 MW of installed capacity. These projects could sell electricity to two of the three major utilities in Malaysia: Tenaga Nasional Berhad (TNB) in peninsular Malaysia or Sabah Electricity Sendirian Berhad (SESB) in Borneo. Project developers had to negotiate a Renewable Energy Power Purchasing Agreement (REPPA) with the relevant utility according to a "willing buyer, willing seller model" and were granted a license for 21 years after the commissioning of a plant. Renewable energy project developers were responsible for the costs of grid connection and utility system reinforcement including cables, transformers, switchgears, protection equipment, and meters, and were required to distribute electricity into the network between 11 to 33 kV. Facilities had to be within ten kilometers of the nearest interconnection point to the grid, with SREP developers paying for grid extension, and all facilities had to meet regulations set by the Ministry of Environment. Lastly, a minimum 30 percent equity had to be held in all projects by *Bumiputera* ("indigenous Malaysian") stakeholders, and foreign companies were allowed to participate only with a maximum equity of 30 percent.

Project developers also had to go through a somewhat cumbersome and complicated process involving an "application for approval" followed by an "application for license." Ancillary support mechanisms, such as Pioneer Status or Income Tax Allowance and tax exemption on equipment, were implemented in tandem with the SREP.[4]

At the time the SREP was launched, it was believed that accomplishing a five percent share of renewable electricity supply by 2005, or a total of 500 megawatts (MW)—the stated goal of the Eighth Malaysian Plan—would save the country about $2 billion over those five years.[5] The SREP was designed to accomplish this as well as multiple secondary goals.

First, it was seen as a way to tap the waste energy potential from the palm oil industry, one of the largest agricultural sectors in the country. One participant commented that "we thought we could get at least 600 MW alone, 100 MW above the [original SREP] target, from the 400 plus palm oil mills producing millions of tons of empty fruit bunches, palm fronds, and palm oil mill effluent each year that currently go to waste." Indeed, at the price of only $30 per barrel of oil, one study estimated the value of palm oil waste at more than $200 *billion*.[6] Given that the price of oil was three times that amount in early 2011, conceivably $600 billion of value exists. One study calculated at least 665 MW of renewable energy

---

4 According to the Malaysia Industrial Development Authority: "Companies undertaking generation of energy using biomass, hydropower (not exceeding 10 MW) and solar power that are renewable and environmentally friendly are eligible for the following incentives: (i) Pioneer Status with a tax exemption of 100 percent on statutory income for 10 years. Accumulated losses and unabsorbed capital allowances incurred during the pioneer period by companies whose pioneer status will expire on and after 1 October 2005 are allowed to be carried forward and deducted against post-pioneer income of a business relating to the same promoted activity or promoted product; or (ii) Investment Tax Allowance of 100 percent on the qualifying capital expenditure incurred within a period of 5 years. This allowance can be offset against 100 percent of the statutory income for each year of assessment. Any unutilized allowances can be carried forward to subsequent years until the whole amount has been fully utilized

These incentives are for applications received from 1 October 2005 until 31 December 2010. Companies which have been granted approval for these incentives but have not implemented the projects are also eligible for these incentives. Companies must implement their projects within one year from the date of approval. For the purpose of this incentive, 'biomass sources' refer to palm oil mill/estate waste, rice mill waste, sugar cane mill waste, timber/sawmill waste, paper recycling mill waste, municipal waste and biogas (from landfill, palm oil mill effluent (POME), animal waste and others), while energy forms refer to electricity, steam, chilled water, and heat."

5 See Mohamed, A.R. and Lee, K.T. 2006. Energy for Sustainable Development in Malaysia: Energy Policy and Alternative Energy, *Energy Policy* 34, 2388–2397. Numbers have been updated to $2010.

6 Shigeoka, H. 2004. *Overview of International Renewable Energy Policies and Comparison with Malaysia's Domestic Policy*. Kuala Lumpur: Pusat Tenaga Malaysia.

capacity from biomass as well, figures presented in Table 11.2.[7] Other studies have calculated a whopping 2,400 MW of potential, with 2,059 MW of this potential coming from the 71.3 million tons of empty fruit bunch produced each year along with 19 million tons of crop residue.[8]

Table 11.2   Biomass electricity potential in Malaysia

	Quantity (kton/yr)	Potential generation (GWh)	Potential capacity (MW)
Rice mills	424	263	30
Wood industry	2,177	598	68
Palm oil mills	17,980	3,197	365
Bagasse	300	218	25
Palm Oil Mill Effluent	31,500	1,587	177
Total	72,962	5,863	665

Second, the SREP was heralded as a way to promote innovation and technological learning in alternatives Malaysia had little experience with, such as waste incineration, small-scale hydro, and solar PV panels. Malaysians produce roughly 20,000 tons of waste every day, enough to "bury the Petronas Towers under a pile of trash every four days," but also enough to create "$10 billion of revenue if converted to electricity." Solar energy potential was cited as "extremely favorable" with 6.0 to 6.5 kWh of potential energy per square meter, given Malaysia's location close to the equator.

Third, the SREP was seen as a mechanism to help achieve the country's remaining electrification goals. While more than 99 percent of the country's population had access to the existing grid in 2000, about 150,000 to 200,000 homes, mostly in the poorest and most rural parts of Malaysia, still relied on diesel generators or received no modern energy services at all. The SREP program was hoped to "develop smaller scale systems, especially mini-hydro and solar, which would reach hard-to-access populations." Another respondent commented that "hydro and solar provide a convenient and cost-effective way to produce power in rural areas as it is near impossible to build transmission lines to cater for the small

---

7   Ibrahim, H. 2002. Small Renewable Energy Power Program for the Promotion of Renewable Energy Power Generation, presentation to the *First Meeting of ASEM Green IPPs Network, Bangkok, October 24–25.*

8   Lim, C.H., Salleh, E., and Jones, P. 2006. Renewable Energy Policy and Initiatives in Malaysia. *International Journal on Sustainable Tropical Design Research and Practice* 1(1) (December), 33–40.

number of homes currently off-grid, and the SREP was believed to help develop suitable off-grid and micro-grid technologies."

Fourth and finally, the SREP was conceived as a way to reduce Malaysia's GHG emissions profile and environmental pollution, especially from the palm oil industry. Every single ton of palm oil, one of Malaysia's largest exports and a stable of their economy, creates 6 tons of palm fronds, 5 tons of empty fruit bunches (EFB), 1 ton of palm trunks, 1 ton of mesocarp fiber, 750 kilograms of palm kernel cake and endocarp, and a staggering 100 tons of palm oil mill effluent (POME). Before it is discharged, POME is usually collected in open ponds or storage takes to degrade, a practice that produces voluminous amounts of methane.[9] Taken together, such emissions from the palm oil industry account for roughly 12 percent of national GHG emissions, and capturing them converts what is in essence a fugitive emission into a source of electricity.

Despite these reasons in favor of renewable energy, however, implementation did not proceed as planned. In 2003, an independent, peer-reviewed study noted that major obstacles remained in Malaysia two years after the SREP had started, including lack of economies of scale, poor perception of commercial viability for projects, and higher risk premiums for financing[10]. At the end of 2005, the SREP had achieved a meager 2.4 percent of its original goal. As the Minister of Energy, Tun Dr. Lim Keng Yaik, noted at the time:

> The SREP ... has not been able to connect the envisaged 500 MW of electricity generated from renewable sources to the national grid. What it has been able to deliver in the last four years is 12 MW from two projects. The wide gap between policy and implementation clearly indicates that there are barriers to the effective transition from a conventional to a sustainable model of energy development.[11]

The Minister's comments were confirmed by an independent study which noted that renewable energy technology was not sufficiently developed in Malaysia.[12] Costs of production were still higher than many other countries at 7 to 25 US cents/kWh, compared to conventional electricity costs of 4 to 6 US cents/kWh. Lack of information on renewable energy was referenced as a major barrier, palm oil facilities were still not converting their waste to electricity, and weak public awareness about the benefits of renewable energy was widespread.

---

9   Sovacool, B.K. and Drupady, I.M. 2011. Innovation in the Malaysian Waste-to-Energy Sector: Applications with Global Potential. *Electricity Journal* 24(5) (June), 29–41.

10   Jaafar, M.Z., Wong, H.K., Kamaruddin, N. 2003. Greener Energy Solutions for a Sustainable Future: Issues and Challenges for Malaysia. *Energy Policy* 31, 1061–1072.

11   Lim, K.Y. 2005. Renewable Energy and Malaysia, Presentation to the *Regional Forum on Sustainable Energy*, Marriot Hotel Putrajaya, April 11.

12   Mohamed, A.R. and Keat, T.L. 2006. Energy for Sustainable Development in Malaysia: Energy Policy and Alternative Energy. *Energy Policy* 34, 2388–2397.

Because of these problems, Malaysian planners extended the SREP for another five years but scaled down its targets to 350 MW: 300 MW in peninsular Malaysia, and 50 MW for Sabah in Borneo. When, yet again, implementation lagged far behind targets, the SREP was revised at the end of 2006 and again in 2007 to increase tariffs, though this was only for biomass and biogas technologies, not mini-hydro and solar systems, as shown in Table 11.3.

Table 11.3   Revised tariffs under the SREP, 2001–2009

Renewable electricity price	Biomass	Biogas	Mini-hydro	Solar PV
$0.057/kWh (2001)	X	X	X	X
$0.063/kWh (2006)	X	X		
$0.070/kWh (2007)	X	X		

Even the revised tariffs, nonetheless, did not significantly accelerate participation in the program. At the end of 2010, Table 11.4 shows that only 61.7 MW of renewable energy capacity had been built from 11 projects. Of these projects, roughly 80 percent were related to waste and palm oil. As Table 11.5 illustrates, an additional 33 projects with 210.85 MW of capacity were in the pipeline but not yet approved or licensed. Total renewable energy supply, including projects supported by the SREP as well as those from other programs and incentives, was 217 MW, less than a one percent share for the country.[13]

For all intents and purposes, the SREP simply didn't work. As one respondent explained:

> The SREP is not a success. From 2001 to 2008, most of the duration of the program, 50 projects were approved for a capacity of 288 MW, but 40 percent were cancelled, and one-quarter were issued with licenses but never started operating. One-third were not issued with licenses, and only 13 MW was built for the first eight years of the program.

Another participant calculated that "about two-thirds of the projects proposed under the SREP never progressed" and that "even today, the major players are not getting into the renewable energy business ... they just don't want to get involved." Still others commented that "our conclusion is that the SREP is a failure. The government needs to relook at it if they wish to see some success in the near future" and "the SREP experience has been dismal."

---

13   Malek, B.A. 2010. *Renewable Energy Development in Malaysia*, presentation to *EU-Malaysia Cooperation in Green Technology, June 1*.

Table 11.4   Licensed and operational SREP projects (as of February, 2011)

Project developer	Project location	Capacity (MW)	Fuel source
TSH Bioenergy Sdn. Bhd.	Tawau, Sabah	10	Empty fruit bunches (EFB)
Seguntor Bioenergy Sdn. Bhd.	Jalan Seguntor, Sandakan Sabah	10	EFB
Kina Biopower Sdn. Bhd.	Lot 2, Jalan Seguntor, Labuk Road Sandakan Sabah	10	EFB
Esajadi Power Sdn. Bhd.	Sungai Kadamaian, Kundasang, Sabah	2	Mini hydro
Esajadi Power Sdn. Bhd.	Sungai Pangapuyan, Kota Marudu, Sabah	4.5	Mini hydro
Recycle Energy Sdn. Bhd.	Lot 3041 & 3042 Mukim Semenyih Daerah Hulu Langat Selangor	5.5	Municipal waste
MHES Asia Sdn. Bhd.	HS(D) 12572, Lot PT No. 3226, Mukim Serting, Negeri Sembilan	10	EFB
AMDB Perting Hydro Sdn. Bhd	Sg. Perting, Bentong Pahang	4	Mini hydro
Renewable Power Sdn. Bhd.	Sg. Kerling, Selangor	2	Mini hydro
Bell Eco Power Sdn. Bhd.	Parit Ju, Batu Pahat, Johor	1.7	Biogas
Jana Landfill Sdn. Bhd.	Puchong, Selangor	2	Biogas

Table 11.5   SREP projects under construction and Approved (but not yet licensed or operational)

Category	Number of projects	Capacity
Mini-Hydro	9	50.8 MW
Biomass	14	140 MW
Biogas	10	20.05 MW
Total	33	210.85 MW

A slew of recent studies have implied much of the same, with one interviewing key stakeholders in Malaysia and finding that regulators and investors "commonly see renewable energy as immature, exotic, unproven, and risky."[14] Another argued that "utilization of renewable energy [in Malaysia] is still very low."[15] The IEA documented that a lack of standard codes and certification procedures, inadequate training, and mistrust among financiers and investors still remained in Malaysia.[16] Others have argued that limited local experience, improperly designed regulations, and a lack of awareness were impeding the diffusion of renewable energy.[17] Yet another study concluded that "renewable energy in Malaysia is still being generated on a small-scale basis [even though] Malaysia is blessed with abundant resources ... The progress in bringing renewable energy generation into the mainstream has been slow."[18]

Challenges

Why, then, did the SREP fail to meet its targets, catalyze the growth of the renewable energy industry, and overcome the barriers it was intended to prevail against? Challenges were partly *technical*, dealing with actual renewable electricity power plant design and training issues; in part *economic*, including low electricity tariffs, unattractive financing rates, and continued subsidies to fossil fuel producers; and partly *institutional*, involving flaws in program design, resistance, and regulatory failures.

Technical

One major technical obstacle involved developing renewable electricity systems that would work in a Malaysian context. One respondent noted that the palm oil industry, for instance, had "no experience with advanced boiler technology and no understanding of biogas technology, meaning engineers encountered numerous problems related to how to combust EFBs or gasify POME." Thus, "biomass projects had to proceed with a lot of costly trial and error." Small hydroelectric

---

14 Sovacool, B.K. 2010. A Comparative Analysis of Renewable Electricity Support Mechanisms for Southeast Asia. *Energy* 35(4), 1779–1793.

15 Lee, C.L. et al. 2009. A Comparative Study on the Energy Policies in Japan and Malaysia in Fulfilling Their Nations' Obligations Towards the Kyoto Protocol. *Energy Policy* 37, 4771–4778.

16 Olz and Beerepoot 2010.

17 Mustapa, S.I., Leong, Y.P., and Hashim, A.H. 2010. Issues and Challenges of Renewable Energy Development: A Malaysian Experience, presentation to the *PEA-AIT International Conference on Energy and Sustainable Development*, Empress Hotel, Chang Mai, Thailand, June 2–4.

18 Oh et al. 2010, 1245, 1251.

projects had trouble "because of the great fluctuation in water supply between the wet and dry seasons, and were also location dependent, given the ten kilometer restriction set by TNB." Most of the Malaysian landfills "were not designed or well suited to capture methane gas," and "the country had practically no experience with designing and using solar panels to generate commercial electricity, which is why no solar projects were ever sponsored by the SREP."

The Bukit Tagar sanitary landfill gas capture power plant we visited, for instance, took "years" to design and involved geo-synthetic clay liners, high design polyethylene blankets, water treatment facilities, anaerobic closing ponds, gas extraction wells, pumps, blowers, flame arresters, and flare stacks. The Kajang waste incineration plant we visited had to specially design a refuse derived fuel cycle that involved magnetically sorting waste, drying and shredding, splitting and recycling non-combustible items, digesting organic waste, combusting the remaining material, and treating effluent. One official at the Bell Palm Oil biogas facility we visited called designing the digester system as "a nightmare." As one respondent summed it up, "when the SREP kicked off, not many people knew about renewable electricity, so we had to develop capacity all on our own."

Another closely related obstacle was lack of skills and insufficient education, training, and quality assurance. While it is clear that all stakeholders were making their own efforts to build their capacity in renewable energy, one respondent noted that "there was no centralized training institution, no place to learn about how to innovate technology, instead we had to do our research on an ad hoc basis." Another respondent mentioned that although his government-funded institute undertakes research in solar, biogas, waste, and other renewable energy technologies, their findings have contributed little toward the SREP, and vice versa.

Such lack of capacity became apparent during two of the site visits we conducted to TNB renewable energy demonstration plants. We showed up at the first, a hybrid system at the Langkawi Cable Car consisting of 16 kWp of solar panels and two 50 kW diesel generators shown in Figure 11.2, only to find that it was no longer working, despite being built in 2002, nine years earlier. As one of the operators explained, "all the solar panels are broken, we think that lightning short circuited the system, we can't get spare parts and wouldn't even know what to do with them if we did."

The second, a TNB solar-wind hybrid project on Pulau Perhentian in Terengganu, was no longer operational due to "lack of maintenance skills" and "interest" from its operators. As one local community member commented, "that system hasn't been working for years even though it's still featured on the TNB website. It's just decoration now, it's been long abandoned."

Even at one of the operational SREP sites we visited, one of the facility managers stated that "spare parts and maintenance is a big problem, we had to hire a full-time technician from China to live at the plant, because there was no one available to train us in what to do." One dimension to this issue of capacity is "poor project feasibility assessments," with "many substandard projects passing the approval stage that never should have." The reason is that "planners didn't really

Figure 11.2   The inoperable TNB solar-hybrid demonstration plant in Langkawi, Malaysia

have the expertise to approve a project, but they did anyways because they wanted to be seen as cooperating with the new SREP." Clearly, the lack of a strong capacity building component within the SREP is a missed opportunity for stakeholders to build expertise, share their experiences, and propagate best practices.

A final technical obstacle relates to onerous interconnection and feasibility requirements stipulated by TNB with significant economic repercussions, since costs had to be borne by SREP developers. One such developer exclaimed that "we were forever at the mercy of TNB to say which substation to interconnect to, what type of circuit breakers and equipment we had to use, how much we had to spend on feasibility studies." That same project developer estimated that such complications accounted to a staggering 20 percent of the total project cost (though other project managers stated that interconnection costs amounted to only 2 to 5 percent of total costs). Another remarked that "it's hard connecting to TNB's grid. There is the legitimate concern that our distribution system could damage their network, but they also set very stringent protection requirements that really complicated the technical efficiency of our project."

Economic

In the economic realm, insufficient tariffs for renewable electricity providers "hobbled" project developers. One respondent noted that "the tariffs paid to SREP developers were not based on sound economic principles, they were set with no consideration of actual cost recovery." The original tariff of $0.056 per kWh was "shockingly low, far below even the actual cost of operating SREP facilities."

In the case of the Bukit Tagar sanitary landfill, respondents commented that it would have made more sense to just use the electricity onsite for the facility, since they were paying $0.10 per kWh for electricity but could only sell at $0.07 per kWh. This, however, "was not allowed," forcing the landfill in essence to "sell electricity to TNB for $0.07 per unit that we then buy right back at $0.10." Another project developer stated that "SREP tariffs pay only enough to cover operation costs and to run the plant, we make no money at all unless we get carbon credits under the CDM." At another facility, participants argued that "the tariff [under SREP] is too low, we don't even get enough to cover operations. We need extra income from tipping fees, recycling, and making plastic resin onsite." For solar projects, respondents mentioned that at least $0.56 per kWh would be needed to make projects viable, "more than seven times the rate currently offered by TNB." As another respondent noted, "the SREP rate is way too low for [solar] projects to be commercially profitable."

Because the tariff was so low, respondents noted that many project developers could make more money doing "other things" with renewable fuels. Palm oil millers, for example, could utilize waste and residues to generate grid electricity under the SREP, or as a component of mattresses and chipboards, in the paper and pulp industry, to make animal feed, or to manufacture compost fertilizer. "SREP is just another option for mills," one respondent mentioned, "but it's not a key factor, something extra, sometimes a nuisance, sometimes worth doing." Landfill gas, similarly, can be used to generate electricity as part of SREP, or when upgraded and "sweetened" used for a variety of other applications including heating, cooking, and as a transport fuel.

A separate economic obstacle dealt with lack of financing and the unfamiliarity of Malaysian banks with renewable electricity projects. One respondent noted that "when the SREP started, local banks were unwilling and unready to give financing. Project developers usually had to go abroad to Chinese and Japanese financiers." Another commented that "bank managers had no idea about renewable energy, it's hard for them to visualize what a landfill gas capture or [a] municipal solid waste plant looks like."

Lastly, subsidies to natural gas and oil, along with energy prices that do not reflect full costs, resulted in an oversupply of electricity generated from fossil fuel sources and impeded the diffusion of technologies under the SREP. As one respondent put it, "the marketplace in Malaysia is not fair. Fossil fuels have been cross-subsidized for decades, eroding the motive to go into renewables ... the playing field isn't only uneven, it's an entirely different game." Another argued that the explanation behind the SREP's poor performance is that "other items and damages associated with fossil fuels, such as carbon dioxide or acid rain, are not factored into tariffs, Malaysians are not paying the full cost of electricity." Such sentiments have been confirmed by other studies, with one commenting that in Malaysia "there is still massive support for conventional energy sources in the forms of subsidies and export credits."[19] The result is a "lack of economies of scale for renewable energy," and "artificially low prices" for fossil fuel supply.

Political and Institutional

Just as significant as technical and economic barriers, a collection of institutional obstacles "wreaked havoc" on the SREP. Our respondents identified no less than seven. First is that the capacity cap of 10 MW was set too low, according to some respondents, for it "ruled out economies of scale" and created a "no man's land" where "smaller projects had too many transaction costs such as interconnection fees and negotiating with TNB, but larger projects were functionally excluded." Yet another jokingly called the SREP "the small-small renewable energy program."

Second, the initial five percent and 500 MW by 2005 national target was set "without any feasibility studies" and "chosen almost randomly, without consultation with key stakeholders in the industry." One participant said that they "didn't know where either of the numbers came from, it surely wasn't by consulting the experts."

Third, altering the tariffs in 2005 was seen as "picking winners" since it only applied to biomass and palm oil technologies, and not solar and hydro technologies. One respondent called the low tariff for solar PV "odd" since that technology had the highest costs compared to all qualified systems. The SREP also functionally picked wind energy as a "loser" since it was excluded from eligibility.

Fourth, REPPAs took "too long to negotiate" and "TNB had the ability to stall or delay whenever they wanted to." Moreover, the 21-year operating license was "hard to meet given that many of the fuel contracts and financing agreements for things like fruit bunches or waste were done on ten- and fifteen-year bases."

Fifth, all SREP projects had to be approved by SCORE, but the committee met only twice a year, meaning if a project missed the first session they would have to wait another six months to apply.

Sixth, neither SCORE nor the Malaysian Energy Commission "had the authority to enforce the SREP, or to reconcile complaints about TNB, or to expedite projects." SCORE and the Commission were also "compromised" and prone to "conflicts of interest" since their members included TNB and the Malaysian Palm Oil Board, but not solar and hydro developers or consumer advocate groups. One respondent went so far as to suggest that "the Energy Commission really does nothing, it's just a surrogate for TNB and has their interests in mind."

Seventh, the SREP had "limited oversight" and "poor evaluation," meaning problems like those above were "not caught or remedied."

Add all of these design flaws up, and some projects took "five years or longer" to get completed, "some developers never bothered to complete the process," and "many more never bothered to start in the first place." Another project developer noted that "it took a terribly long time to go through the SREP process—it was

---

19   Oh et al. 2010, 1251.

supposed to be done in three months, and took us ten times longer." Furthermore, the ten-kilometer restriction to the grid meant that many rural areas and *Orang Asli* communities were still too far away to qualify for a project.

Another key institutional problem was resistance from TNB, the only buyer in peninsular Malaysia that renewable power producers could sell to. TNB had "all the bargaining power," "didn't need renewable energy," and "saw the SREP as a threat to its revenue and profits." "It was a recipe for disaster," another respondent told us, "with TNB setting extremely low tariffs, setting performance provisions that facilities had to deliver electricity at precise capacity factors, levying penalties for shortfalls in expected generation, demanding half of the savings developers accrued from tax reductions, intentionally delaying the approval of projects to put pressure on developers to accept unfair provisions, and setting unfair standby tariffs for backup power."

All three of the site visits we undertook revealed that project developers had asked to install larger systems and sell greater amounts of electricity but had been denied from TNB: the Kajang facility could generate 10 MW but was limited to exporting less than 7 MW; Bukit Tagar shown in Figure 11.3 could do 6 MW but was limited to 1 MW; Bell could have done 4 MW but was limited to 1.7 MW. One respondent explained that:

> TNB sees independent renewable power production as a lose-lose situation because it displaces their capacity, and lowers their electricity sales. Putting them in charge of the SREP was akin to letting a fox manage a chicken coop, or an atheist in charge of a church. TNB put up hurdles every way they could.

Figure 11.3   The 1 MW Bukit Tagar landfill gas capture plant, Malaysia

A final institutional challenge related to the lack of a national, cohesive, strongly implemented policy framework on renewable energy. As one respondent put it, "SREP was designed and implemented with no thought or relationship to the various other ongoing renewable energy programs. There was no harmonization, no coordination." The Ministry of Energy, for example, had to "compete" with the Malaysian Energy Commission and TNB over the "direction of national electricity policy" in addition to "actors like the Ministry of Housing taking charge of waste, the Ministry of Science technological development, the Ministry of Environment regulations, it's a convoluted policy landscape." Another respondent spoke about a "mishmash between the SREP and local policies and regulations that would still impede projects even after TNB would give approval, with hydroelectric projects, because they deal with water, and waste projects, because they interfere with tipping fees, especially polemic." Apparently communication between local and national governments was "poor" and requirements and standards "differed state to state."

Conclusion

Even though the SREP did not meet its targets, it offers insight for energy planners and policymakers. At the top of the list is better design: SREP was "crippled" from the start by capacity caps, a lengthy approval process, explicit picking of biomass and biogas as "winners" but wind and solar as "losers," lack of monitoring, exclusion of stakeholders, and few (if any) pre-feasibility studies. Project developers had to pay the cost of interconnection and had to build systems within ten kilometers of the existing electricity grid. Operating licenses were stipulated to be 21 years but financing agreements and fuel contracts rarely extended beyond ten years. Electricity tariffs were changed under the program in 2001, 2006, and 2007, and targets were revised downward in the middle of the program. Projects supposed to take three months to design, approve, install, and connect ended up taking three to five years, and scores of project developers abandoned their efforts midstream. Each of these design flaws, especially the 10 MW cap, artificially, and perhaps unnecessarily, prevented an organic renewable electricity market from taking root.

In addition, the efficacy of the SREP was eroded by fragmentation and lack of cohesion with other Malaysian energy policies, notably continued subsidies for natural gas and oil as well as conflicts with state guidelines and policies concerning hydroelectricity, waste-to-energy, and palm oil effluent and waste. A sort of policy gap existed between the lofty targets enshrined in the SREP and the local developers and officials on the ground charged with realizing those targets.

Relying on the dominant state-owned electric utility TNB also proved to be a mistake, as was the "willing seller, willing buyer" model of REPPAs. Rather than embrace renewable energy, evidence from our pool of respondents strongly suggests that TNB opposed it and used a variety of tactics, such as interconnection fees, costly feasibility studies, and delays to discourage projects. Part of this is understandable, given that the structure of the SREP meant that small-scale renewable electricity projects traded off with TNB revenues and profits.

Equally important, electricity tariffs under the SREP did not match true production costs, were not based on sound economics, and did not provide cost recovery for project developers. Every single project we visited highlighted the need for extra streams of income, from CDM credits to tipping fees and recycling, in order to be financially viable. When tariffs under the SREP were changed, this was made without costing studies and created the perception of yet again "picking winners" of biomass and biogas. Ultimately, these diffuse factors took a severe toll on the overall performance of the SREP.

Chapter 12

The Teachers Solar Lighting Project in Papua New Guinea

Introduction

Papua New Guinea (PNG) represents one country where energy poverty is acute. Roughly 80 percent of its 6.3 million inhabitants subsist in rural communities far away from the capital and each other, and 90 percent of the entire population lacks access to the national electricity grid.[1] Most communities are days away from the nearest transit point to even initiate lengthy journeys to urban areas, and the country reports the lowest electricity access level in the entire Pacific.[2]

Small-scale, off-grid renewable energy technologies such as SHSs offer villages in Papua New Guinea and elsewhere the ability to generate electricity and provide reliable energy services. This chapter explores one national program there intended to distribute SHS to rural schoolteachers known as the Teacher's Solar Lighting Project, or TSLP. With support from international donors, the $3 million TSLP was designed to disseminate SHSs to 2,500 teachers, eventually distribute 14,500 units, and jumpstart a self-sustaining local market for solar equipment. By the project's close, however, only one SHS constituting 50 Watts-peak (Wp) of capacity had actually been sold—less than one ten thousandth of the original target.

This chapter explores why. It analyzes the design and implementation of the TSLP before moving into the reasons it failed to achieve its targets. Unlike other chapters, this section has none on benefits, given that the program distributed only one unit. Since the TSLP did not meet its targets, it offers value as a "worst practice" example for development and energy specialists. Our analysis also refuses to explain the failure of the TSLP in purely political and economic terms, looking at technical, social, and cultural dimensions as well.

---

1   Nagai, Y., Yamamoto, H., and Yamaji, K. 2010. Constructing Low Emitting Power Systems through Grid Extension in Papua New Guinea (PNG) with Rural Electrification. *Energy* 35, 2309–2316; World Bank 2010a. *Reducing the Risk of Disasters and Climate Variability in the Pacific Islands: Papua New Guinea Country Assessment.* Washington, DC: World Bank Group.

2   World Bank 2005. *Request for GEF Funding: Teacher's Solar Lighting Project.* Washington, DC: World Bank Group, Project ID P088940.

Description and Background

PNG relies significantly on imported energy fuels. PNG, along with other small island developing states such as Kiribati, Vanuatu, Samoa, the Solomon Islands, Fiji, and East Timor, is greatly dependent on energy imports. This dependence makes it susceptible both to global price spikes in the cost of these fuels as well as possible interruptions in energy supply. Moreover, dependency on fossil fuels, particularly oil, results in severe macroeconomic shocks.[3] Collectively these factors make off-grid sources of energy such as kerosene much more expensive over the long-term than alternatives such as SHSs. Planners in Papua New Guinea therefore designed the TSLP as a logical extension of a previous program called "Solar Lighting for Rural Schools" (SLRS), supported by Japanese aid donors, which spent about $13 million distributing solar kits to 320 primary schools in 1997 and 1998 to minimize their dependence on kerosene for lighting (Department of Education 2000). By its close, that program successfully installed 1,688 solar systems spread across 2,400 classrooms providing energy services to 75,000 children in 89 school districts (approximately 15 percent of the rural school age population), but it suffered from two problems. First, it placed lighting systems in classrooms but not in the actual teacher's residences. As one interview respondent explained:

> The SLRS was intended to raise the quality of schools and facilities, not the domiciles of teachers. During its design phase it was proposed that some of the lights might need to be taken out of the classroom from the existing schools, and connected to teacher's homes in close proximity. This idea was rejected on the grounds that the SLRS was to assist communities, not individuals. Basically, the project was focused on lighting up the school, but keeping the teachers in the dark.

Second, virtually none of the thousands of SHSs installed under the SLRS remained operational five years after the program ended, with "only a few still working in 2003" and "no ongoing after sales service or maintenance."

The germination of the TSLP thus began in the early 2000s as a way to make the lives of rural teachers "easier," with most "unable to afford solar equipment due to modest salaries." Most of these teachers endure what respondents described as "hard lives" involving isolated and difficult to reach communities, inconsistent paychecks, frequent reassignment, and expensive lighting. As one respondent elaborated:

> Most teachers in PNG instruct without light. If they are lucky enough to have a kerosene lantern, they may use it for a few hours a week, but then it's costly, produces poor quality light, is rough on the eyes, and presents a fire and safety hazard.

---

3 Bacon, R. and Kojima, M. 2008. *Vulnerability to Oil Price Increases: A Decomposition Analysis of 161 Countries.* Washington, DC: World Bank, Extractive Industries for Development Series.

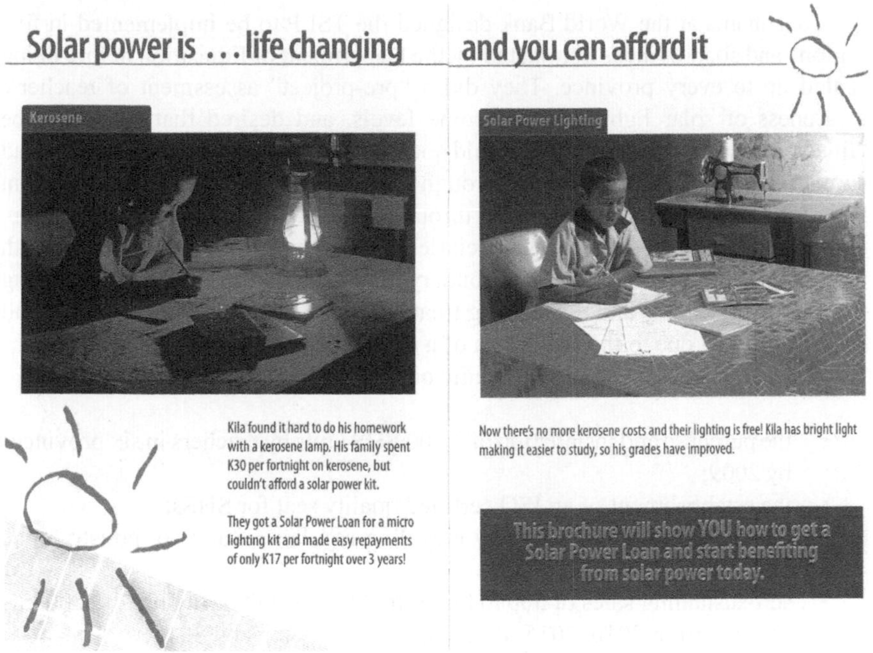

Figure 12.1   An advertisement for SHSs in Papua New Guinea

One independent assessment calculated in countries such as PNG, classroom light levels are as low as two percent of western standards and that teachers frequently grade homework with light levels one percent of western standards.[4] SHSs, by contrast, provide teachers and students with high quality light, displace costly and polluting kerosene, and enable teachers to prepare lectures and grade after dark, benefits espoused in advertisements like the one shown in Figure 12.1.

With $2,942,600 in support from the GEF, the World Bank, the PNG Sustainable Development Program, and the national government, the TSLP kicked off in 2003. Project documents state that its main objective was to "improve the life of rural human services providers by making available affordable, environmentally sound, basic electricity services from renewable energy." Stated goals were to improve teacher retention and health and establish a robust local market for solar equipment. The TSLP was broken down into three components: a financing package intended to make SHSs affordable for the roughly 50,000 teachers in PNG; a technology improvement component requiring manufacturers to obtain certification and regularly produce a catalogue of certified solar components; and a consumer awareness component focused on "extensive outreach" to SHS purchasers.

---

4   Mills, E. 2006. *Alternatives to Fuel-based Lighting in Rural Areas of Developing Countries*. Berkeley: Lawrence Berkeley National Laboratory.

Consultants at the World Bank designed the TSLP to be implemented in five regions and six provinces nominated by the Department of Education before being scaled up to every province. They did a "pre-project" assessment of teacher's awareness of solar lighting and income levels, and desired that the TSLP be "financially sustainable" in that it would not give subsidies to purchasers but instead support "financial intermediaries" through "loan risk equalization." Confidence in solar technology was to be improved through accreditation, product specification, and product training. Primary beneficiaries were to be teachers as well as health workers posted to remote areas; secondary beneficiaries were children attending rural primary schools and those being treated in rural health clinics, as well as all citizens of PNG due to the scaling up of a renewable energy market.

Project documents stated six specific quantitative goals to be achieved:

- the purchase and installation of 2,500 SHS units by teachers in six provinces by 2009;
- the establishment of an ISO certified quality seal for SHSs;
- the implementation of battery recycling regulations and an industry code of conduct;
- self-sustaining sales of 500 SHSs from 2005 to 2009 growing to 2,000 kits per year from 2010–2015 (or a total of 14,500 units by 2015);
- reduced kerosene consumption by ten to twenty liters per household per month through SHS adoption; and
- about 250,000 tons of carbon dioxide equivalent displaced over the course of the program.[5]

In the end, however, the TSLP met few of these targets. One respondent lamented that "only one unit was ever sold, and I don't even know if it was to a teacher." Even the World Bank's own evaluation noted that:

> With only one solar PV loan (processed in October 2008) the project has not achieved its target of 2,500 systems and the key outcome has not been met ... Despite the project's relevance to the country, a need to encourage the market for supply of solar PV equipment to teachers in rural areas, and a strong demand, the project did not meet its objectives. A failure to fully address the commercial realities of the market and inadequate supervision justify an overall moderately unsatisfactory rating.[6]

---

5   Global Environment Facility 2005. *Medium-Sized Project Proposal Request for GEF Funding: Teacher's Solar Lighting Project*, Washington, D.C.: GEF; World Bank 2005. *Request for GEF Funding: Teacher's Solar Lighting Project*. Washington, DC: World Bank Group, Project ID P088940.

6   World Bank 2010b. *Implementation Completion Memorandum (ICM) for GEF MSP-Papua New Guinea Teachers Solar Lighting Project*. Washington, DC: World Bank Group.

Ultimately, as another respondent bluntly declared, "The TSLP sucked down a few million bucks, not a single teacher benefitted, and all participants really got for their years of participation in the project were headaches and perhaps free tea and biscuits during the various meetings we held trying to figure out what was going on."

What in the world, then, happened?

Challenges

This section categorizes the obstacles facing the TSLP into technical, economic, social, and political dimensions. Technical barriers related to equipment problems and a dearth of eligible systems. Economic ones included high rates of poverty and lack of financing, as well as the selection of participating suppliers based on a least cost approach rather than a quality of technology approach. Social challenges encompassed expectations about what SHSs could offer as well as issues of jealousy, vandalism, and theft. Political barriers reflected poor institutional capacity and the *wantok* system of "big men."

Technical Factors

Technical obstacles include logistical and equipment challenges and lack of product diversity.

As one commercial distributor of solar systems explained, "trying to acquire high quality panels or batteries in PNG is a procurement nightmare." Even when high quality panels were available and desired by customers or local bank branches, getting them where they needed to go proved difficult. PNG is the largest of the Pacific Island states, with 463,000 square kilometers spread across 600 islands and atolls.[7] Transport is "close to impossible" with the road network "reaching less than one percent of the country." Everything, from food and automobiles to people and solar panels, must be flown around the country's 492 airports. This "tyranny of terrain makes distributing SHS to the remote, rural population difficult," as panels often break or fall apart in transport, or suffer delays and damage from natural disasters.

However, an even greater barrier is related to a severe shortage of eligible products. Although the TSLP did result in higher quality panels being certified by the International Electrotechnical Corporation as well as some ISO certified suppliers, only one type of system—a 50 Wp panel with 3 lamps, a charge controller, a lead acid battery, a power box, cables and accessories shown in Figure 12.2—was chosen for the program, even though PNG suppliers offered a variety of other models and applications. The SHS for the TSLP could only offer lighting; it could not produce electricity for radios, televisions, mobile phones, or other

---

7   World Bank 2010.

PNG Teachers Savings and Loan Society Limited

PROJECT SPONSORS
Global Environment Facility
PNG Sustainable Development Program Limited
PNG Government

PROJECT IMPLEMENTING AGENCY
PNG Sustainable Energy Limited

Figure 12.2   **The TSLP catalogue featuring a 50 Wp solar home lighting kit in Papua New Guinea**

devices. The TSLP, as one participant commented, "offered only one product with no customer flexibility, and merely decided what people needed without asking them first."

Economic Factors

Economic barriers involved cultural conceptions of savings or money, the high up-front cost of a SHS, a reliance on participating financial institutions and suppliers that lacked robust rural networks, and systems that did not match the true energy needs of users.

One substantial impediment related to what one participant referred to as a complete "lack of financial literacy," with many teachers in rural areas possessing "no sense of savings, credit, debt, or even money." As one consultant involved SHS distribution explained:

> Many ordinary people in other developing countries will invest in a SHS and talk about fuel savings, payback, and even discount rates, concepts we in the West take for granted, but in PNG people are completely unfamiliar with them. People have no concept of the future, let alone money. In many tribes, no word for "next week" or "future" even exists. People are not used to understanding time as something that passes or the idea of "saving," so asking them to calculate the net savings from a solar system versus kerosene discounted into the future, or to compare capital investment to interest and savings, is impossible.

Another participant remarked that most ordinary teachers have "no understanding of investment or discount rates or interest, so one key flaw of the TSLP was its dependence on financial institutions to disseminate SHSs through loans."

For those teachers financially literate enough to express interest in the TSLP, the upfront cost of the system (more than 3,000 kina, or $1,390) was still beyond what they were able to afford (with "typical" salaries ranging from 600 to 800 kina, or $278 to $371 per month). This meant that teachers would have to save for "months" or even "years" before they could participate in the TSLP, and as one solar manufacturer in PNG told us, "most purchasers are looking at upfront costs only." One principal calculated that if teachers were frugal, they can realistically save about 1,000 kina, or $463 per year, meaning they would need to save for three years before they could buy their own SHS. Another principal commented that "many teachers have children, and the cost of a SHS loan is approximately equivalent to the cost of a year of private school, which makes solar an unwanted and unnecessary luxury."

Flaws also became apparent in how the TSLP selected participating financial institutions and suppliers of SHSs. Financial institutions had to meet eligibility criteria that functionally excluded many microfinance and provincial banks as well as smaller lending institutions. Respondents told the author that "the big multinational banks were not interested in the TSLP" since it suggested interest rates on loans at 12 percent per year compared to a "normal" rate of 24 to 36 percent per year. This left only "one bank in the middle," the Teachers Savings and Loan (TS&L), which seemed on the surface a logical choice since they already marketed financial services to teachers.

However, this reliance on the TS&L as the sole financier proved unwise. Most teachers "were not part of the TS&L and didn't want to join," preferring instead to receive their paychecks in cash. Moreover, the TS&L set "strict" loan requirements such as teachers needing continuous payroll deduction, a minimum of 1,000 kina in the account, a record of "credit worthiness," and stipulations that they could not have any other types of loans, in addition to other criteria. This proved problematic when teachers had outstanding debts at other institutions for things like automobiles, weddings, and schooling.

Furthermore, respondents iterated that the TS&L relied on a "convoluted loan and financing structure." The TS&L offered only three-year loans for SHSs, but the TSLP dictated five-year loans, which meant lenders ended up creating a two tiered loan system where the World Bank paid two years of the loan up front and the teachers were required to pay off the rest of the loan over three years, with the TS&L underwriting the risk. This created two problems: most teachers didn't want "the hassle" of trying to meet loan requirements they saw as "complicated," and the TS&L was reluctant to fully promote the program out of fear they would be left with significant costs should teachers not actually purchase a large number of systems.

In addition, neither the pilot provinces nor the suppliers of the SHSs were chosen "on solid economic grounds." Instead, provinces were selected "politically," with one respondent suggesting that Bougainville was included only since it was recovering from a civil war. Many of the selected provinces did not have extensive TS&L locations, creating a mismatch between financial networks and provincial coverage. Another participant noted that "there were actually many teachers from other provinces who were interested in the TSLP, but were ineligible because they happened to live in the wrong place."

Similarly, SHS manufacturers and suppliers were apparently selected "based on low cost" instead of their "ability to make and distribute high quality SHS units." Based on a "competitive tendering process," only two suppliers could participate in the TSLP: Westlink Enterprises and Roots Electrical. Interestingly, those with extensive existing networks, and backed by Nationwide Bank and ANZ, such as Active Power Systems in Port Morseby; ESCO in Port Moresby, Goroka, and Madang; G4S Communications, Rural Power Supplies, and Tolec Electronics, were excluded. As Table 12.1 shows, one of these two chosen suppliers, Roots Electrical, didn't even have offices in six TSLP provinces and the other, West Link Enterprises, had only one location per province, and three of these retail outlets did not have addresses or phone numbers, just post office boxes, implying they were not yet fully established when West Link made their bid.

This de facto emphasis on cost rather than quality proved disastrous. To win the tendering process, respondents informed us that both suppliers "bid too low."

Table 12.1   Supplier networks of SHSs for the TSLP

Province	Roots Electrical	West Link Enterprises
National Capital District	Boroko	Port Moresby
Milne Bay	-	Alotau
Western Highlands	-	Mount Hagen
Autonomous Region of Bougainville	-	Buka
Western Province	-	Kiunga
New Ireland	-	Kevieng
East Sepik	-	Wewak

Roots Electrical presumably underbid to the degree that they had problems with collateral, ended up pulling out of the project in 2007, and went bankrupt in 2008. Through the course of their tenure, they never brought a single SHS into PNG. West Link in good faith bought 200 SHSs, but ended up "having to pay for them all" after teachers failed to approach TS&L for loans, even though PNG Sustainable Energy apparently "promised they would cover any risk." West Link then spent 2005 to 2009 litigating in court to try and recover their investment, creating what one respondent called "a complete farce rather than a dedicated effort to grow the SHS market." Another secondary distributor, who purchased systems from West Link, also said he "almost went bankrupt" due to the TSLP and "would have if he hadn't limited his purchase to a small number of units."

A final economic impediment concerns the inability of SHS to meet the energy needs of teachers. The TSLP did a pre-feasibility study of solar awareness and income levels, but never actually "asked teachers what they wanted." This was a "major mistake," with respondents stating that "no matter where it comes from, what teachers want is light—they don't care if it comes from solar or hydro, a flashlight or a kerosene lamp." Kerosene, for example, is "well established as a fuel of choice" in rural areas as a cash sale item with roughly $60 million in annual sales. It is widely seen as portable and reliable; one respondent noted that "kerosene is sold in coke and beer bottles distributed through a readily existing network, people are comfortable with it." By focusing on technology (a SHS) instead of the desired energy service (lighting), the TSLP "unfairly narrowed its promotion efforts."

Relatedly, many teachers wanted "more than just lighting." As one respondent put it:

> Local rural villagers and teachers need electricity beyond its ability to give them light, they are now interested in mobile phones, mp3 players, DVD players, refrigerators, computers, and other devices, and need more than a 50 Wp panel to keep them going.

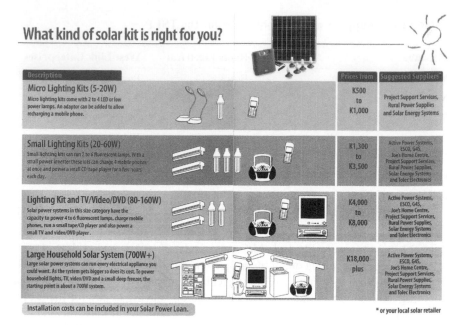

Note: As of January 2012, 1 PNG kina was about $0.43.

Figure 12.3   Range of Solar Energy Equipment Available in Rural Papua New Guinea

Far better investments for these teachers could be any one of the systems shown in Figure 12.3, capable of meeting comparatively large demands for energy, or smaller solar cell phone chargers and lanterns (sold at only 90 kina, or $42) which could be hooked on a wall or carried around like a flashlight or torch, also sold by a variety of vendors. These technologies were either "lighter and more portable than a clunky SHS," or "big enough to provide all of a home's energy needs." A 50 Wp system, by contrast, fell in a sort of "no man's land" between these various classes of technologies.

Also, teachers are "nomadic" in the sense that they do not always stay in the same school or house for long. As one respondent clarified:

> Most teachers in PNG stay for only one year ... It was a mistake for the TSLP to target teachers, instead it should have targeted schools or the Managing Board of school districts, who would keep panels in one place. A teacher that doesn't know where they will be next year will not invest in a SHS that might be useless if they're moving to an area with electricity access, or where kerosene might be significantly cheaper.

Making investments in a SHS was thus perceived as "risky" since it would likely have to be relocated wherever the teacher would need to go.

Social Factors

Social challenges to the TSLP revolve around cultural biases, theft and vandalism, and low consumer awareness.

Local rural homes in PNG, for example, are not built well for a SHS because they have sack sag roofs, making it difficult to properly mount the unit. Teachers and many rural villagers, moreover, spend very little of their time inside their house. As one respondent explained:

> SHS work well for individualistic households, but PNG is not an individualistic society. It is very communal, and people spend most of their days outside or interacting with the community, meaning a fixed wire solar system doesn't make sense. A rural house is essentially used only for sleeping; the rest of the time people are attending social gatherings outside under trees or throughout the neighborhood, not inside the home. Put another way, solar energy is good for Westernized permanent houses, but 95 percent of the country does not live like that, and even the 5 percent that do rarely spend enough time inside the home to justify investing in a SHS.

A related problem is that living with a SHS necessitates a shift in lifestyle. As one participant noted:

> One has to learn to live with solar energy, adjusting to a sort of solar lifestyle. On sunny days a user can live life similar to that in a Western house with television, phones, and computers, but that's only if the weather permits and if their system is fully charged. Otherwise one has to make hard choices about whether they want lights or radio for a few hours, rather than both. Maintaining a SHS also requires diligence. Given the expensiveness of lead acid batteries, most PNG solar users rely on used car batteries which need to be maintained once a month: checked for water, cleaned, and Vaseline placed on charges. Solar is not a technology one can install and forget about, one needs trained to use it, it is not as easy as Western experts who've never actually lived with the technology think it is.

A second challenge relates to an abnormally high frequency of theft and vandalism of solar panels within PNG distributed by other programs, due largely to jealously and sabotage, creating a reluctance to commit to the technology. As one respondent stated:

> One of the big problems with SHSs in PNG is that they have been promoted at the household or community level, but such things—houses and communities— don't really exist in the same way they do in other countries. In PNG there is

no such thing as a geographic community; we instead have clans, tribes, and family groups that can extend within and beyond geographic communities. An "individual" or "communal" SHS is not likely to work one year later, because it is seen as belonging to everyone and therefore to no one. No one sees themselves as responsible for it.

Introducing a solar panel into PNG homes can assault their communal value system, as it benefits only one household or community rather than the entire clan. Tribal communities can therefore react aggressively and negatively. As one respondent put it, "distributing solar panels gives rise to intense jealousy. People think that because they do not have one, no one else should, so they break and destroy all of the ones that they come across." Another respondent commented that "a solar panel cannot be shared, cannot be distributed communally, so it does not fit with the core values, the cultural mores of PNG rural life. It is therefore determined to be offensive, and sometimes destroyed or removed as a result."

These responses may strike some readers as extreme, but when we visited the Talidig Primary School in the Sumkar District of Madang Province, previously home to an 85 Wp system, upon arrival we learned that only a few days earlier the entire thing, save the pole-mounted panel, had been stolen from the classroom. The author also encountered a vendor selling a SHS likely stolen from an elementary school (its configuration matched panels given under the SLRS, but not sold commercially in PNG) when driving through the Daulo Pass, seen in Figure 12.4.

A third social barrier to the TSLP involved lack of information and knowledge among teachers about SHSs. One respondent commented that "the very factors that make teachers need solar energy—their remoteness, their isolation—also makes it difficult to get them information about solar technology." Indeed, none of the many teachers, principals, and schools we visited had ever heard of the TSLP. TS&L's strategy of "information awareness" also proved inadequate; it was limited to one demonstration computer touch screen system in one office in each of the six provinces participating in the program, but this proved "ridiculously expensive and ineffective," since few teachers actually visited these TS&L offices regularly. A far more efficacious campaign, one respondent noted, should have involved printed advertisements, road shows, demonstrations in villages, and seminars at national educational events, as well as discussions on a Department of Education radio show broadcast twice a week called "Teacher's Tea Time." Promotional efforts for the TSLP, moreover, had to compete with a well-publicized campaign for solar loans from competitors of the TS&L.

This lack of knowledge promulgates inaccurate expectations about what a SHS can accomplish. Some villagers we met with thought a single 10 Wp system could chill and freeze large amounts of food, another that it could power all appliances in a home, yet another that it could run three air conditioners. Rural communities in PNG have what one respondent called "a big misunderstanding of solar electricity." We also visited with teachers that excitedly discussed about how they believed that

Figure 12.4 A vendor selling an ostensibly stolen solar panel near Daulo Pass in Papua New Guinea

a small solar panel could produce enough energy to power a computer laboratory and copy machine, or a kitchen with an oven, microwave, and a bread maker.

Political Factors

A final class of political challenges revolves around institutional problems with TSLP program managers, the *wantok* system of patronage, and an inability to collect feedback and learn from past failures.

Institutionally, rapid turnover within the World Bank and PNG Sustainable Energy—management of the TSLP switched three times in one year—essentially meant that responsibility "shunted from one person to the next like a hot bête nut." Delays in implementation took almost two years before the TSLP received its financing, and then managers did not spend all of the money disbursed, only about ten percent in the first year by one contractor's estimate. One participant commented that "infighting and bickering caused the TSLP to collapse."

Management at the programmatic level was described by one respondent as a real "basket case," in part because of "stringent, rigid, and bureaucratic" requirements set by international donors, but also due to a "personality clash" between managers at the World Bank, the designer, and PNG Sustainable Energy, the implementer, who apparently by the end of the TSLP disliked each other "strongly enough that they refused to sit in the same room."

Additionally, the concept of *wantok*, meaning "one nation," served as an impediment. This idea, "widely held by rural households, has resulted in everyone believing they are a relative to everyone else." *Wantok* means people are not willing to ask for money for a service that they think they are entitled to. Very few people in villages, for instance, may actually pay for kerosene or electricity themselves. Instead they "rely on tribal leaders or politicians, or a 'big man,' someone who is wealthy, to pay for things directly or give them money." In this type of an economy, SHSs are seen as gifts rather than commodities.

Thirdly, TSLP managers appeared not to learn from or incorporate feedback from similar schemes or past failures in PNG. One participant said that the "standard response from the GEF and World Bank when things go wrong with their energy aid programs is to ignore them." Another stated that such institutions prefer to "manage energy projects from behind the desk of their air conditioned offices, and avoid getting out into the rugged bush to understand energy poverty or energy problems on the ground." Yet another argued that:

> Institutions like the World Bank or USAID think that if you have a good idea or new energy technology, you are 90 percent there, and implementation takes the remaining 10 percent. Experience here suggests it's really the opposite: 10 percent the idea, and 90 percent the training, the consumer awareness, the promotion. Getting the technology right is completely secondary to effective promotion in gaining social acceptance.

More specifically, a national solar water pumping project, the SLRS, a Ministry of Petroleum project, and a South Pacific Institute of Renewable Energy project all tried to distribute solar technology throughout PNG in the 1990s. Each of these projects encountered "barriers similar to the TSLP" and also resulted in "hundreds of failed systems, improperly installed, no longer working, abandoned and forgotten"—emphasizing the need for maintenance and information awareness, and problems related to vandalism and theft—yet such lessons were not infused into the TSLP. One respondent noted that "when the TSLP was being designed, they didn't even consult with managers from any of these four programs—they just believed that they knew best." Indeed, in February 2010, when the research team decided to select the TSLP as a case study, the World Bank's website still listed it as a "success," implying that the institution has difficulty acknowledging failures.

Conclusion

The TSLP expended millions of dollars in PNG in an attempt to distribute thousands of SHSs, failed to meet its targets, and resulted in the confirmed installation of only one solar panel. The reasons for its poor performance transcend technical, economic, social, and political boundaries. The difficulty of getting high quality panels and components manufactured and distributed in PNG and a lack of product diversity constrained consumer choices. Poor financial literacy, high capital costs, badly chosen participating financial institutions and suppliers, and a mismatch between SHSs and energy needs further impeded progress. Cultural norms, theft and vandalism, low public awareness, lack of institutional capacity, a belief that energy services should be free, and an inability to learn from past mistakes compounded these impediments.

However, the "worst practice" nature of the TSLP ironically implies what a "best practice" approach might have entailed. Asking what teachers want, or how end users would use electricity, rather than presuming what they need would have immediately shifted the design of the TSLP away from a single type of system—a 50 Wp SHS—to a variety of technologies and devices including portable solar chargers, solar lanterns, and higher capacity solar home systems. Choosing financial institutions and suppliers with existing networks in selected provinces instead of those that would have to build new networks, and selecting them based not on a "least cost tendering" process but a "high quality service" approach, would have minimized delays and potentially avoided bankruptcies. Placing responsibility for solar loans with the schools or managing boards of schools, rather than the teachers, would have reduced risk and better accommodated the actual lifestyles of teachers, who often change districts and assignments. An educational and awareness campaign consisting of active demonstration, road shows, and radio broadcasts would have been more effective than a passive computer demonstration at a limited number of TS&L branches. An energy service company "fee for service" model, where each school board could have paid a small fee to hire one full time person responsible for servicing and maintaining all solar panels, batteries, and parts within a district, could have minimized potential problems with maintenance and vandalism.

The tribulations of the TSLP highlight the necessity of having a robust and complete policy process of design, implementation, distribution and financing, monitoring and evaluation, and use. The World Bank and GEF relied on consultants, concept papers, and appraisals but did not do due diligence in designing the TSLP to match the energy needs of its targeted users. The implementer, PNG Sustainable Energy, clashed with the World Bank, had delays in procuring equipment, was sued by the TSLP's supplier West Link Enterprises, and suffered rapid turnovers in personnel. One of two suppliers went bankrupt, the other had only proxy offices in half of the participating provinces and spent most of its time litigating against PNG Sustainable Energy, and neither understood how to undertake extensive marketing to raise consumer awareness about solar energy. The single financier,

TS&L, had limited outreach into rural areas and a convoluted loan structure that oddly excluded most of its intended users. Designers and implementers were apparently so occupied in dealing with delays and personality conflicts that they did not institutionalize feedback and evaluation sufficient to correct problems when they arose. Teachers remained uninformed about solar energy and reluctant to invest a significant proportion of their income in a solar loan to the degree that only one—and possibly none—ever participated in the TSLP. These facets suggest that a successful program needs *all* of the steps of the policy process in place in order to succeed.

Chapter 13
Lessons Learned

This chapter presents 12 broader lessons for energy policymakers, development practitioners, scholars, and even students regarding what our case studies teach us about energy poverty and renewable energy. It presents our 12 lessons, drawing from successes and failures, in no particular order, though the evidence we've collected suggests that all of them are important. Table 13.1 provides an overview of the 12 interconnected lessons and 42 factors attributable to successful programs; the inverse of the 42 factors correlate roughly with our failures. Some of these factors are closely related to each other and overlap, others cut across multiple lessons. Elements of having strong technical standards and certification, for example, appear in both "Appropriate Technology" and "Evaluation and Monitoring." Similarly, elements of focusing on energy services rather than technology appear in "Appropriate Technology" and "Flexibility."

Net Beneficial Energy Access

Perhaps the simplest lesson is that programs that invest in renewable energy systems, or expand access to modern energy, can spill over into enhanced development and the achievement of higher living standards, lowered fuel consumption or fuel prices, improved technology, better public health, and reduced greenhouse gas emissions. Most impressively, they tend to provide these benefits with a positive cost benefit curve; that is, their benefits exceed their costs. Each of our six successful case studies confirmed this lesson.

GS in Bangladesh has helped more than one million people, mostly women and children, acquire access to cleaner, cheaper, better quality energy services and created a robust local renewable energy market with thousands of jobs. There, relatively small things such as better meals, radios, light bulbs, and telephones make a huge difference in individual lifestyle and enjoyment. It has also reduced national greenhouse gas emissions by hundreds of thousands of tons per year, a reduction of 15 percent for the entire electricity sector.

China's REDP saw the incomes and standards of living among its participants, mostly rural nomadic herders, rise significantly through their use of small-scale wind turbines and SHSs. SHS use has specifically resulted in such improvements as increased family communication levels, increased workable hours and access to information through radio and television. The use of more primitive and inefficient fuel sources such as ghee and kerosene was also found to have decreased due to SHS usage. We discovered that nomadic herders specifically appreciated the benefits of

Table 13.1 Twelve lessons and 42 factors associated with our case studies

Lesson	Factor	Bangladesh	China	Laos	Mongolia	Nepal	Sri Lanka	India	Indonesia	Malaysia	Papua New Guinea
Net beneficial energy access	Expanded access to energy services	X	X	X	X	X	X	X	X	X	
	Job creation	X	X	X	X	X	X	X	X	X	X
	Lowered fuel consumption/prices	X	X			X	X				
	Improved technological quality	X	X			X	X				X
	Reduced morbidity and mortality			X			X			X	
	Fewer greenhouse gas emissions	X			X	X					
Appropriate technology	Feasibility studies	X	X	X	X	X	X				
	Scaling up	X	X				X				
	Service rather than technology orientation	X	X	X	X	X	X				
	Technical standards and certification		X		X		X				
	Cultural sensitivity	X	X	X	X	X	X				
Community commitment	Community ownership/operation/participation	X	X	X	X	X	X				
	Minority/gender empowerment	X		X		X					
	Monetary contributions (cash, savings, collateral)	X	X	X	X						
	Non-monetary contributions (time, labor, land, materials)					X	X				
Awareness raising	Marketing and promotion	X	X	X	X	X	X	X		X	X
	Demonstration	X	X	X	X	X	X				
After-sales service	Product guarantees/warranties/buy back	X		X	X						
	Training/funds for maintenance	X	X	X	X	X	X				

Income generation	Classes in productive end-use	X	X	X	X	X	X		X	X
	Scholarships	X								
Institutional diversity	Involvement of non-state-actors/private sector	X	X	X	X	X	X	X	X	X
	Polycentricity	X	X	X	X	X	X			
	Cost sharing	X	X	X	X	X	X			
	Avoidance of corruption				X	X				
Affordability	Provision of credit/microcredit/ESCO "fee-for-service" model	X	X	X	X	X	X	X	X	X
	Revenue collection	X		X	X	X	X			
	Support for manufacturing/industry		X	X	X					
	Lower programmatic costs	X		X	X	X				
Capacity building	Institution building	X	X	X	X	X	X			
	Outsourcing		X	X	X					
	Improved business practices (accounting, auditing, revenue collection, marketing)			X	X	X				
	Self-sufficiency	X	X	X	X	X	X			
Flexibility	Diversity of eligible technologies	X	X	X	X	X	X			
	Follow-up project		X	X	X	X	X			
	Promotion of both grid/off-grid systems			X	X	X				
	Adjusted targets/extended deadline			X	X	X	X	X	X	X
Evaluation and monitoring	Independent evaluator	X	X	X	X	X	X	X	X	X
	Penalties for noncompliance	X	X	X	X	X	X			
Political support	Policy integration	X	X	X	X	X	X		X	
	Dedicated or experienced implementing agency	X	X	X	X	X	X			X
	Project champion/political leadership	X	X	X	X	X	X			

Note to Table 13.1: An 'X' signifies that a program meets a particular factor

SHS through lighting and electricity for mechanical milk separators. Importantly for some, SHS technology was necessary to maintain their nomadic needs in the summer months, demonstrating that modernity can go hand-in-hand with tradition.

In Laos, a combination of grid-connected hydroelectricity and off-grid SHSs installed during the REP has reduced the drudgery of collecting fuelwood and the dangerous health impacts from indoor air pollution. Apart from for lighting and entertainment purposes, electricity generated through the REP has increasingly been used for productive uses. Particularly in grid-connected areas, households are able to use electricity for agriculture, refrigeration, ironing, and small businesses such as restaurants and convenience shops.

In Mongolia, the REAP enabled herders to receive energy services during the harsh winter months, primarily electricity from portable SHSs for lighting, mobile phones, satellite television, radio, and cooking. This meant they used significantly less coal and had an improved carbon footprint. The program additionally displaced almost 200,000 tons of carbon dioxide equivalent emissions.

In Nepal, evaluations of the REDP have specifically documented as much as $8 in benefits per household for every $1.40 expended. The program saw the technology cost for microhydro systems significantly decline, in some cases by more than 50 percent.

In Sri Lanka, roughly three times the budge of the ESD--$150 million—was invested in the renewable energy market from 1998 to 2004, implying it catalyzed the involvement of the private sector. The ESD allowed end-users to take charge of their electricity needs in the absence of government provision, reaching villages which would have otherwise not have received electricity services. It improved the financial status of many rural villages and resulted in productive activities such as sewing and carpentry; and also substantially lessened consumption of kerosene, producing public health benefits. Most importantly for rural Sri Lankans, the ESD procured access to television and studying at night, seemingly two of the most important benefits of electricity.

Even our four failures support our contention about net beneficial access: the VESP in India allowed the rural population to enjoy more efficient and cost-effective energy sources for cooking and lighting, as did the Indonesia SHS Project. The SREP in Malaysia did slightly reduce air pollution and lower greenhouse gas emissions. The TSLP in Papua New Guinea ultimately improved technological quality by (unintentionally) bankrupting poorly performing distributors.

The common formula for each of these cases is that as programs distribute more and more renewable energy systems, technology costs decline, capacity factors improve, incomes rise, and community wellbeing advances. The qualitative benefits of renewable energy partnerships and programs like these are backed by scores of quantitative assessments that reach similar conclusions.

One scientific study simulated what it would cost to provide universal access to gaseous fuels for cooking and electricity for lighting in India by 2030, and found (again) that program benefits would far outweigh expenses. Improved living standards, livelihood opportunities and climate change mitigation—just three benefits—more than justified the cost of expanding energy access.[1] Another study looked at the benefit cost ratio from 2005 to 2015 for switching away from fuelwood, dung, and coal in 11 developing countries to cleaner forms such as improved cookstoves.[2] Such efforts would cost only $650 million to achieve, but would produce $105 billion in benefits each year.

A third study found that mechanized services such as grinding and milling enabled women to increase their agricultural activities by an average of 33 percent.[3] Another study comparing off-grid renewables in China with conventional gas and diesel generators in Xinjiang, Qinghai, and Inner Mongolia concluded that "in all three provinces, renewable energy systems are economically superior to conventional energy options, even before social and environmental benefits are included."[4] Analogously, a comprehensive economic assessment of solar and wind systems concluded that they were lower than the costs of centralized diesel generation.[5]

Similarly, in Brazil's northeast state Ceará, access to modern electricity coincided with the largest improvement in any state of Brazil's Human Development Index ranking. As one assessment noted when analyzing this case study, "energy services are undeniably tied to economic development at the macro level."[6] Another has concluded that "in modern times no country has managed to substantially reduce poverty without greatly increasing the use of energy or utilizing efficient form of energy and/or energy services."[7]

The ADB recently surveyed the promise of renewable energy and expanding access in the Asia-Pacific region, and noted that its ability to save labor and time among rural communities was significant. The study argued that renewable energy access not only meets basic needs, but also diversifies and expands sources of income, increasing both the scope of local employment and the status of skill

1   Reddy, B.S., Balachandra, P., and Nathan, H.S.K. 2009. Universalization of Access to Modern Energy Services in Indian Households—Economic and Policy Analysis. *Energy Policy* 37, 4645–4657.

2   Hutton, G., Rehfuess, E., and Tediosi, F. 2008. Evaluation of the Costs and Benefits of Household Energy and Health Interventions, presentation to the *Clean Cooking Fuels & Technologies Workshop, June 16–17*, Istanbul, Turkey.

3   Porcaro and Takada 2005.

4   Byrne, J. et al. 2007. Evaluating the Potential of Small-Scale Renewable Energy Options to Meet Rural Livelihoods Needs: A GIS- and Lifecycle Cost-Based Assessment of Western China's Options. *Energy Policy* 35, 4391–4401.

5   Thiam, D-R. 2010. Renewable Decentralized in Developing Countries: Appraisal from Microgrids Project in Senegal. *Renewable Energy* 35(8), 1615–1623.

6   Thiam, D-R. 2010.

7   P. Sharath Chandra Rao, Jeffrey B. Miller, Young Doo Wang, John B. Byrne, "Energy-microfinance intervention for below poverty line households in India," *Energy Policy* 37 (2009) 1694–1712

development and training. Moreover, expanded access can improve the resilience of poor communities to handle other social and economic setbacks such as natural disasters or the closure of local factories, enabling them to prosper rather than merely survive. The study found that the unit cost of energy decreases as one moves up the energy ladder, related largely to improvements in efficiency as well as the reduced effort required for energy access, freeing up time and income for other needs and aspirations. Lastly, it noted that high-speed transportation, telecommunications, information technology, and a variety of things that make life better depend on electricity; none of them can function on traditional fuels. As the study concluded, "denying people access to modern energy is thus equivalent to depriving them of the fruits of human economic and technological progress in virtually all fields. Instead, it confines them and their future generations to a low-yield, labor intensive life and denies them the means and tools to raise incomes and escape perpetual poverty."[8]

For these reasons, investments in small-scale renewable energy technologies in developing countries pay dividends well beyond their original costs.

Appropriate Technology and Scale

Successful programs frequently start small with pilot programs or with feasibility studies before initiating full-scale projects and scaling up to greater production or distribution volumes. They almost always choose appropriate technologies matched in quality and scale to the energy service desired. They set technical standards so only high-quality systems enter the marketplace, and they often possess culturally sensitive dissemination programs.

Feasibility studies and piloting are useful ways to identify market segments and determine if enough demand exists for renewable energy systems. The REDP in China, REP in Laos, REDP in Nepal, REAP in Mongolia, and ESD in Sri Lanka all relied on World Bank-sponsored feasibility assessments well before they were implemented, and GS conducted willingness to pay studies of consumers in Bangladesh to determine if they would pay more for solar energy than kerosene or dry cell batteries. For the REP in Laos, at least half the households in a given village had to express an interest in renewable energy before the scheme would extend to them. Correspondingly, in Nepal only communities expressing an interest and desire for energy participated in microhydro schemes, and they had to know how to build, own, repair, and manage a microhydro unit before they were given one.

In terms of scale, GS started in Dhaka before expanding go the rest of Bangladesh, the ESD in two Sri Lankan provinces before expanding to eight, and

---

8   Jamil Masud, Diwesh Sharan, and Bindu N. Lohani, *Energy for All: Addressing the Energy, Environment, and Poverty Nexus in Asia* (Manila: Asian Development Bank, April, 2007)

the China REDP initially targeted six provinces in the Northwest before expanding to another three—demonstrating the utility of scaling up.

Successful programs also have an orientation towards energy services, matched in quality to end-uses, rather than technological deployment; they recognize technology not as an end itself, but a gateway to a particular energy service. The lack of this flexibility also featured prominently in our failures: the Malaysian SREP, Indonesian SHS project, and TSLP in Papua New Guinea all presumed to know what consumers wanted and restricted them to one or a select few eligible technologies. By contrast, the different approaches undertaken by our successful case studies certainly reflects the multi-dimensionality of energy use and did not confine or limit how technologies (and applications) could be utilized. For example, one respondent in Bangladesh informed us that GS:

> Does not, like other groups, mistakenly think of 'energy services' as a single entity; there are really several types of demand for energy, as variegated as a patched quilt: heat energy comes from cooking from firewood, and its uses vary from boiling water to drying yarn and making bricks, SHS can do everything from light study areas and charge mobile phones to power DVD players and televisions, biogas units provide cooking as well as fertilizer for farming. The key here is flexibility and a pairing of energy supply with energy quality and household needs.

The REDP in Nepal and ESD in Sri Lanka relied on simple technologies matched in proper scale to the communities they were intended to serve. Rather than build new grids, communities were encouraged to construct simple microhydro units in the range of 10 kW to 100 kW to be owned and operated by communities themselves. Program managers at the REDP empowered communities to adapt systems to meet their own needs. In Dhading, where mustard does not grow, mechanical energy is used to husk rice. In Kavre, more agriculturally oriented, grinding and mustard seed expelling are given priority. In Lukla, a tourist destination home to hotels, tourist facilities consume most of the electricity. Both programs also had strong environmental standards so that microhydro units did not damage rivers or degrade forests. Similarly, GS designed each of their programs in Bangladesh to meet distinct energy needs: ICS for the "poorest," SHS for slightly more affluent rural households, biogas systems for wealthy rural households or communities. In Mongolia and China, herders put SHSs to use separating milk (see Figure 13.1), and in Sri Lanka, charging mobile phones.

Even so, it is still a challenge to get it "exactly right" among successful programs. During our site visits in Laos, some respondents from fishing villages mentioned that they would have preferred grid instead of SHSs so that they could refrigerate fish. Similarly, in Sri Lanka, some respondents talked about changing expectations and energy needs in targeted communities over time. While households were initially happy to have access to SHSs or village hydro systems to power black-and-white TV sets and basic lighting, there is now increasing

Figure 13.1   A solar-powered milk separator in China

demand for more energy needed to operate blenders, refrigerators, irons, and other modern electrical appliances.

The point here is that different classes of people will put energy to use in different ways. In their work on energy poverty, Shonali Pachauri and Daniel Spreng identified numerous sub-groups of people:

- tribal people and leaders
- rural unemployed
- landless peasants
- traditional merchants and craftspersons
- peasants with small farms
- the urban poor
- servants
- blue collar employees.[9]

For these dissimilar classes of people, the services that energy delivers are far more important than the carrier itself. What matters is the relative importance

9   Pachauri, S. and Spreng, D. 2008. Some Remarks on the Choice and Use of Indicators of Development, Presentation to the *Clean Cooking Fuels & Technologies Workshop, June 16–17, Istanbul, Turkey.*

users attach to it.[10] A successful renewable energy project must not only consider the appropriateness of the technology diffused for the needs of the targeted communities but also anticipate how these needs will change and grow over time, and facilitate the transition to larger wattages or different technologies. As Amory Lovins mused many decades ago, "people do not want electricity or oil, nor such economic abstractions as 'residential services', but rather comfortable rooms, light, vehicular motion, food, tables, and other real things."[11]

Furthermore, programs that work tend to promote or harmonize rigorous technical standards to ensure renewable energy technologies perform as expected. This underscores the reliability component of energy access, and it also serves as a meaningful form of consumer protection. As one of our interviewees argued:

> People will pay for energy services, just not for unreliability or unpredictability; they won't pay for electricity that is on when they don't need it or off when they do need it. Nor will they pay for electricity that has such erratic fluctuations in voltage that it fries appliances—that's what they don't want to pay for. But reliable, efficient service—yes, they want that.

Thus, successful cases strengthened technology in tandem with institutions and community awareness.

China's REDP, for example, focused on whole-cycle quality improvement for solar panels and SHS. It executed a "start-to-finish" quality process by establishing manufacturing standards and practices, facilitating access to product certification, and introducing a randomized testing regime which penalized companies at the production-line and retail stages for non-compliance with system performance requirements. It also culminated in the "Golden Sun" label to certify compliance with REDP's standards. REAP in Mongolia established technical standards and procedures for testing the quality of SHS and WTS devices and mandated that only qualified systems could receive support under the program. SHS were even sent to China for testing to ensure compliance. Standardization and certification have been proven to facilitate more widespread manufacturing, reduce costs, and improve quality of systems in other studies.[12] In Sri Lanka, the ESD mandated that technologies meet national standards and technical compliance had to be verified by chartered engineers.

---

10 Van Der Vleuten, F., Stam, N., and, Van Der Plas, R. 2007. Putting Solar Home System Programs into Perspective: What Lessons Are Relevant? *Energy Policy* 35, 1439–1451.

11 Lovins, A.B. 1976. Energy Strategy: The Road not Taken. *Foreign Affairs* 55(1), 65.

12 Barnes, D.F. and Floor, W.M. 1996. Rural Energy in Developing Countries: A Challenge for Economic Development. *Annual Review of Energy and Environment* 21, 497–530.

Effective programs are sensitive to cultural differences. In Mongolia, because nomadic herders are always on the move, the REAP supported portable SHSs that could be easily transported, assembled, and disassembled. In Nepal, Sri Lanka and rural China, largely agrarian communities, microhydro units fit in nicely with the cultural mores of farming communities that needed some combination of mechanical power, irrigation, milk separation, and electricity. In Bangladesh, women technicians have been trained at Grameen Technology Centers and repair and inspect SHS because men are not traditionally permitted to enter domiciles while heads of households are working during the day.

Our four failures, by contrast, demonstrate the inverse: few feasibility assessments before design and implementation, pushing preselected technologies, lacking technical standards, and failing to appreciate culture. The SREP in Malaysia conducted assessments of the technical potential of renewable electricity before it began, but did not consult manufacturers or end-users. Malaysian planners seemed obsessed with achieving targets, not delivering services, to the extent that much of the electricity from SREP projects actually flows past un-electrified consumers to distribute their energy to the commercial grid.

In both the Indonesian and Papua New Guinean cases, planners pushed solar energy because expert consultants told them to, not because communities had expressed an organic demand for the technology. While a pre-feasibility study of solar awareness and income levels was undertaken in Papua New Guinea, it did not actually "ask teachers what they wanted". Moreover, SHSs were practically an assault on the local culture of Papua New Guineans, who did not spend significant time inside their homes, consequently making a fixed SHS "useless." Households there viewed solar energy as an affront to their communitarian culture, did not have modern conceptions of money and time (mooting the efficacy of the program's financial model based on credit), and considered energy a public good provided under a *wantok* system rather than a commodity.

In Indonesia, the dismal sales of SHS units were not simply a reflection of reduced purchasing power of end-users and retailers due to the Asian Financial Crisis. Competing government initiatives deploying SHS through grant mechanisms severely hampered the market-based approach of the program. While end-users did not mind using SHS units, they naturally preferred waiting for government subsidies or buying from the cheaper second-hand market. In this regard, the selection of provinces was based on a presumed rather than expressed need from the communities targeted. While the program did successfully develop technical specifications, testing, and certification, the inexperienced and disinterested BPPT was not able to translate these gains into opportunities for other stakeholders, especially for fledging local SHS companies, most of which went bankrupt.

In India, although the promotion of biomass gasifiers, biogas systems, and improved cookstoves was supposed to leverage on the abundance of biomass sources in rural areas, the program did not anticipate the technological and logistical complications of maintaining such systems, challenges exacerbated by a lack of standards, and the reluctance of communities to change their fuel habits.

Households also found it more profitable to sell energy fuels to secondary markets rather than to those wishing to utilize them to generate energy services.

For the SREP, one major obstacle was to develop renewable electricity systems that would work in the Malaysian context. The palm oil industry had very little experience with biogas and biomass technologies. In addition, many landfill sites in Malaysia were not suited for landfill gas capture. Equally troubling was the lack of centralized training and quality assurance, meaning that participants had to research and develop their own technologies with little institutional support.

Community Commitment

Effective programs actively promote community ownership, in-kind contributions of labor, time, and other resources, and participatory decision-making and planning. They tend to target minority groups in rural areas (such as female heads of household or children). They do not "give away" renewable energy technologies or over-subsidize technology or research. In essence, these attributes ensure that households become involved in projects, and that key stakeholders remain active.

Most of our successful case studies had local communities pay for renewable energy systems themselves, and also saw local households or village leaders operate the technology, meaning they become invested in how they perform. The projects had very high payback rates, with 90 to 99 percent of households paying back loans on time.

Having local communities pay for renewable energy projects with their own funds means they express interest and responsibility in how they perform; they become not only passive consumers, but active participants. As one respondent explained, "classically, energy planners have seen the access question as one involving 'givers' and 'takers:' the utility giving electricity or donors giving technology, and the consumers taking it. This completely places the energy services provider and consumer into a false dichotomy, one which successful programs break."

Contributions do not have to be financial, either; communities and households can donate time (such as digging a canal), land (such as free property for the project site), or resources (such as wood for distribution poles). As the World Bank has noted, "participation of local communities, investors, and consumers in the design and delivery of energy services is essential."[13] Successful projects also tended to have inclusive and participatory modes of decision-making.

For instance, in Bangladesh, GS established 45 Grameen Technology Centers to train members of the communities they operate in to maintain and repair their own systems, be it SHSs, ICSs, or biogas digesters. In fact, GTCs prioritize training programs for women which have contributed toward improving their social and

---

13 World Bank, *Rural Energy and Development for Two Billion People* (Washington, DC: World Bank Group, 1996).

economic standing. The Mongolian REAP and Chinese REDP established testing centers closer to communities so that they could be more involved in maintenance. The Laos REP involved villager leaders in the formulation of local electricity companies. Microhydro projects in Nepal and Sri Lanka were designed to accelerate community-level participation by asking for voluntary land donations for the construction of canals, penstocks, power houses and distribution lines, and they were also managed exclusively by home-grown, village based cooperatives and groups. Villagers were also required to contribute labor and civil works. One additional benefit of strong community participation in microhydro projects in Nepal is that the units were not prone to being targeted by Maoist insurgents during the civil war.

In terms of empowerment, GS in Bangladesh recruits marginalized and socially disadvantaged groups, such as women and youths, to train as technicians. For the REP in Laos, a gender sensitive "Power to the Poor" subcomponent targeted female heads of household, offering them interest free loans to help cover interconnection costs to hydroelectric mini-grids. In Nepal, the REDP established separate gender-based community organizations at the village level so that women felt free to participate in microhydro decisions.

Conversely, while some VESP test projects, especially those run by NGOs, emerged as vehicles to encourage greater community involvement, in many more, communities were only involved nominally and continued to expect or rely on government officials to operate and manage the systems. Additionally, 90 percent of project costs were provided through grants from the government and 10 percent provided by villagers either in-kind or in cash.

In Indonesia, little effort was made to fully understand the true energy needs of the targeted population or to involve them in the project design. Price incentives gave the "wrong" signals, with other ongoing programs giving away solar panels for free and the government increasing subsidies for kerosene, lowering its cost and obviating the financial value of investing in a SHS. The national government expanded the grid to communities that had just purchased SHS units, flooding the secondary market and mitigating the desire for households to pay for systems.

Malaysia employed a straight forward top-down approach in terms of selecting eligible technologies, determining tariffs, and adjusting program targets. There is little evidence that local communities were consulted in any of these decisions.

In Papua New Guinea, individual ownership of SHSs is a concept that is not understood and in fact considered an insult to the community, resulting in vandalism and theft of units and components.

Raising Awareness

Successful partnerships do not take consumer awareness or information about renewable energy for granted. They have robust marketing, promotional, and demonstration activities, providing avenues for the public not only to understand

Grameen Technology Center

Grameen Shakti has taken steps to use knowledge on renewable energy technology as a tool for women empowerment, unleashing their social and economic potential. GS has established 45 Grameen Technology Centers (GTCs) which have already trained more than 3,000 women on the repair and maintenance of Solar Home Systems and other electronic equipment. These women work as technicians on an entrepreneur basis to provide repair and maintenance services to SHS owners or produce SHS accessories as demanded. Additionally, 115,000 women from SHS user families have been trained on the proper maintenance and repair of their own systems. GS has also extended outreach to around 10,500 students who have participated in GS Educational Programs, which promote awareness about renewable energy technologies and the environment.

Wind Energy Micro-enterprise Zone

Grameen Shakti is currently conducting research to measure the potential utility of wind energy in the coastal areas of Bangladesh. The present phase of the program will allow Grameen Shakti to gather financial and technological information for possible future expansion on wind energy technology.

Solar Thermal Program

Grameen Shakti is conducting research for the development of high quality solar water heaters, solar cookers and solar driers.

Solar Powered Computer Education Center

Grameen Shakti has successfully developed a solar powered computer education center. GS hopes to establish more of these centers to increase access to computer education in rural areas.

International & National Awards Received

Grameen Shakti has received the following awards for its innovative financing policies and strong after-sales service based on user satisfaction.

GRAMEEN SHAKTI

Grameen Bank Bhaban   Tel : 9004081, 9004314
19th Floor, Mirpur-2   Fax : 8035345
Dhaka-1216, Bangladesh   E-mail : g_shakti@grameen.com
web : www.gshakti.org   g_shakti@grameen.net

Grameen Shakti

Pioneering the Promotion, Development and Outreach of Renewable Energy Technologies to the Rural Poor

- Illuminating Homes with Solar Power
- Facilitating Comfortable Cooking with Biogas
- Encouraging a Healthy Environment
- Increasing Income and Alleviating Poverty
- Improving the Standard of Living

Better life, More income

Figure 13.2   A GS brochure in Bangladesh

the project better, but to also become better educated in renewable energy development.

Marketing and promotion involve the publication and distribution of sales catalogues, informational brochures, product displays, websites, and advertisements (in print, radio, or on television). These can be directly supported by the institution or program itself, like the brochure published by GS in Bangladesh shown in Figure 13.2, or the poster from REDP in China in Figure 13.3, or indirectly supported by participating partners, such as the Sri Lankan company advertisement in Figure 13.4.

Indeed, almost all of our cases—including the failures—had some degree of promotion. Apart from GS in Bangladesh and the ESD in Sri Lanka described above, the REDP in China supported the production of TV and movie content to expand awareness about renewable energy, as well as initiated training capacity-

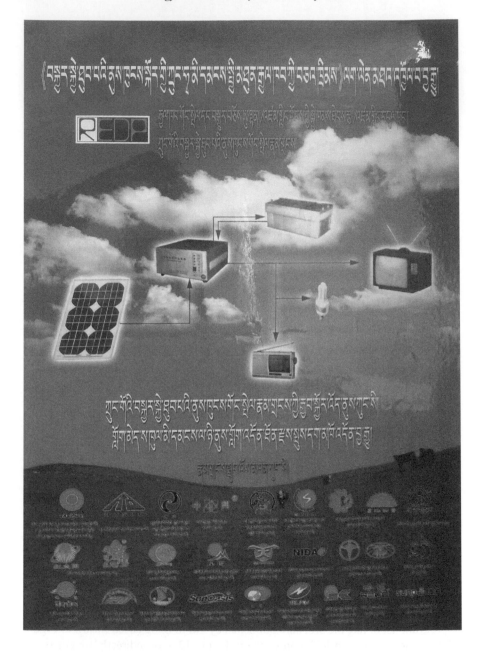

Figure 13.3   A promotional poster from the REDP in China

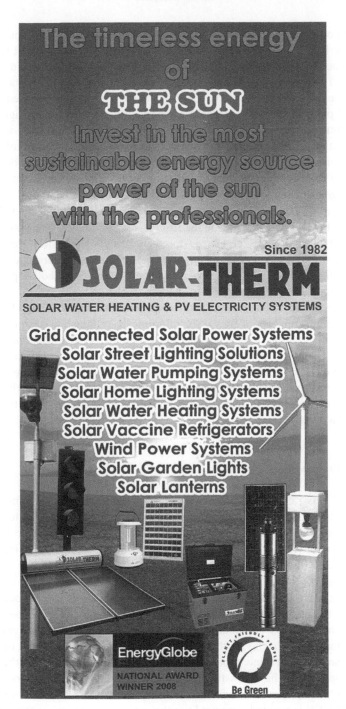

Figure 13.4 A Solar Therm advertisement in Sri Lanka

building courses and conferences for PV companies; the REP in Laos, REAP in Mongolia, REDP in Nepal, VESP in India, SREP in Malaysia, and TSLP in Papua New Guinea had brochures and printed advertisements. The REAP in Mongolia also had a brainstorming workshop involving consultation with stakeholders, donors, banks, and suppliers; the ESD a Business Development Center which promoted public awareness through workshops and marketing campaigns.

Promotional efforts appear most successful when coupled with technology demonstration. In Bangladesh, GTCs conducted large demonstrations of solar and biogas devices and GS employees sometimes embarked on door-to-door visits to familiarize communities with the technology. GS engineers consistently worked with village leaders to distribute brochures, hold science fairs at local elementary schools, and host workshops for policymakers. The REDP in China, REAP in Mongolia, the REDP in Nepal, and the ESD in Sri Lanka did "road shows" where experts travelled around rural areas to display and demonstrate technologies to village leaders and household members. Innovative dealers and financiers involved with the ESD would even visit factories, churches, and other public areas to demonstrate units to large groups of people in one go.

By contrast, in India, local plantation experts and many village households remained uninformed about the VESP. In Indonesia, outreach was negligible, with dealers unable to afford television and radio commercials and even brochures and catalogs. In Malaysia, demonstration was nonexistent if not counterproductive (all of the demonstration facilities we visited were no longer working despite being only a few years old), and in Papua New Guinea awareness raising activities were inappropriately located in bank offices far away from the teachers that needed actual solar systems.

After-sales Service

Successful programs strongly emphasize after-sales service and maintenance, ensuring that technologies are cared for by rural populations or technicians. This can occur on the "supply side" through product guarantees, warranties, and assurances to buy back systems if communities are connected to the grid; or on the "demand side" through training sessions and free maintenance.

For example, once GS technicians in Bangladesh sell and install equipment, they do not leave it up to the consumer to care for it. GS runs a buyback system where clients can return their systems at a reduced price to the organization, and it gives free maintenance and training to all existing clients so that they can care for and maintain their systems by themselves. They teach each user how to properly maintain and conduct minor repairs, and also offer a free warranty for the first few years of operation. They view the needs of households and customers as "never ending." One expert we interviewed commented that "the customer satisfaction achieved by GS is so good compared to other programs that many actually switch from other companies to GS just because of their after sales service." They train hundreds of technicians each year in renewable energy maintenance and the manufacturing of selected components.

China's REDP explicitly devolved maintenance authority to dealers. The REP in Laos ensured that maintenance and battery replacement were formal parts of each program, with responsibility clearly delegated to private sector participants. The REAP in Mongolia gave financial support for after-sales service call centers and the establishment of warranties. It created new centers to help herders maintain their systems, provide advice on battery charging, distribute spare parts, and honor warranties. The REDP in Nepal and ESD in Sri Lanka utilized some funds to hold microhydro maintenance training sessions. In Nepal specifically, a small amount of every electricity tariff enters into a mandatory maintenance fund; in Sri Lanka specifically, the ESD supported the creation of 80 permanent service and distribution centers with $5 million committed from the private sector.

In India, maintenance expenses were not internalized in VEC budgets which led to meaningful differences in the overheads required to keep VESP systems effectively functioning. Because of dismal sales, after-sales maintenance was practically non-existent in both Papua New Guinea and Indonesia. In Indonesia in particular, SHS vendors were generally overwhelmed from having to provide credit lines to potential customers and collect installments to pay enough attention to after-sales services. As a consequence, Indonesian customers had to be responsible for their own maintenance and servicing. Operators in Malaysia were so inadequately prepared for maintenance that they had to either outsource it to Chinese technicians or abandon systems altogether (such as the desertion of both of TNB's renewable energy demonstration projects).

Income Generation

More effective programs couple and cultivate energy services with income generation and employment, they don't just "wait" for it to occur. They also sometimes offer scholarships and university training.

GS in Bangladesh offers a scholarship competition for the children of SHS owners. It sponsors technical degrees in engineering and related fields for employees that commit to staying with the organization long-term. Also, GS has also done an excellent job linking its products and services to other local businesses, and integrating its technologies with other programs. As one example, it connects the use of biogas units in homes and shops with the livestock, poultry, agriculture, and fishery industries. Clients wishing to own their own biogas unit can also purchase livestock, and clients that do not wish to use the fertilizer created as a byproduct from biogas units can sell it to local farmers, aquaculturists, and poultry ranchers. Similar linkages have been made in the promotion of GS's solar panels, mobile telephones, compact fluorescent light bulbs, and light emitting diode devices.

In parallel, China's REDP offered nomadic herders tips on how they could use solar electricity not only for lighting but also to separate milk and cheese, charge mobile phones, and refrigerate yoghurt. In Laos, services put to use by solar panels have increased the business of restaurants, hotels, teahouses, and shops. In Mongolia, improved access to cellular telephony from REAP has enabled herders to get better commodity prices for cashmere, meat, livestock, cheese, milk, yoghurt, and curd. In Nepal, the REDP has linked microhydro energy and the promotion of non-lighting uses of electricity including agro-processing, poultry farming, carpentry workshops, bakeries, ice making, lift irrigation, and water supply. In Sri Lanka, the ESD motivated many homeowners to begin new enterprises such as selling baked goods and vegetables, and existing shop owners to extend their operating hours after dusk. In fact, in the successor RERED project, income-generating activities were made a mandatory component.

One of the objectives of the India VESP was to provide energy for productive purposes, yet since the value of biomass fuel was seen as free, and systems kept breaking down, few concrete opportunities for generating income materialized. In Indonesia, some SHSs were used to power cottage industries, agriculture, fisheries and other small enterprises. However, the overall program did not place any formal emphasis on income-generating activities. In Malaysia, although the overall SREP yielded subpar results, lack of institutional support seemed to have pushed some participants to develop their own income-generating activities in addition to selling electricity to the grid. For example, waste-to-energy projects received revenues from tipping fees, recycling, on-site manufacturing of plastic resin, fertilizer, and carbon credits, in addition to electricity sales. Unfortunately, there was no platform provided to share and improve on these innovations with other stakeholders. In Papua New Guinea, as only one sale was made, no income-generating opportunities occurred.

The key lesson here is that successful programs did not just supply energy or electricity, presuming people know how to use it; they instead teach them how to put that energy to productive use. In essence, these projects succeeded because they promote the types of economic activities that go hand-in-hand with modern energy, enabling communities to form strong livelihood groups, to process agricultural commodities and crops, and to sustain small businesses and enterprises such as bars and restaurants.[14]

Institutional Diversity

Effective programs distribute responsibilities among different institutional partners, involving a diversity of important stakeholders in each project,

---

14   Jooijman-van Dijk, A.L., Clancy, J. 2010. Impacts of Electricity Access to Rural Enterprises in Bolivia, Tanzania, and Vietnam. *Energy for Sustainable Development* 14(10 (March), 14–21.

especially "non-state actors."[15] This allows for the sharing of risks as well as organizational multiplicity which can create "checks and balances" on other actors involved in the project. Typically, these actors transcend multiple scales, making them "polycentric."[16] Successful programs always involve a degree of cost sharing between governmental or intergovernmental entities, the private sector, and communities. Sometimes, they even intentionally avoid corrupt levels of government.

Each successful case involved heterogeneity of institutions at a variety of scales, meeting our first two criteria of diversity and polycentricity. Rather than run things from Dhaka, GS in Bangladesh now has a network of more than 1,000 offices spread throughout the country. GS enrolls communities into renewable energy projects at the household and village level but also engages district and national policymakers along with international donors and lending firms. China's REDP involved actors at not only the global scale (World Bank and GEF) but the national scale (NDRC), provincial scale (governments), and corporate scale (solar and wind manufacturers). The REP in Laos divided tasks so that grid-expansion was carried out by EdL on a cross subsidization model but off-grid SHS were diffused according to an ESCO model involving Village Off-grid Promotion and Support (VOPS) offices, Provincial Electricity Supply Companies (PESCOs), village electricity managers (VEMs), and policymakers at the Provincial Departments of Energy and Mines. The REAP in Mongolia enrolled the World Bank and GEF along with the National Renewable Energy Center, Ministry of Fuel and Energy, Ministry of Finance, Ministry of Environment and Resources, and local *soum* centers.

The REDP in Nepal is especially diverse and polycentric, for it collaboratively works with various layers of community organizations, including governments, and social networks. The REDP partnered with and built on earlier work from the UNDP, Danida, USAID, and the ADB. In the REDP, District Development Committees and District Energy and Environment Sections share experiences institutionalizing bottom up participation in the project. National actors like the AEPC and Ministry of Environment, Science, and Technology, as well as global actors like the World Bank and UNDP, play significant roles as well.

---

15   van Dorresteijn, D.B. 2011. Partnerships with Non State Actors to Improve Local Service Delivery to Achieve the MDGs, presentation to the United Nations Economic and Social Commission for Asia and the Pacific (UNESCAP) and International Fund for Agricultural Development (IFAD) Inception Workshop on *"Leveraging Pro-Poor Public-Private-Partnerships (5Ps) for Rural Development,"* United Nations Convention Center, Bangkok, Thailand, September 26.

16   See Ostrom, E. 2010. Polycentric Systems for Coping with Collective Action and Global Environmental Change. *Global Environmental Change* 20, 550–557; and Sovacool, B.K. 2011. An International Comparison of Four Polycentric Approaches to Climate and Energy Governance. *Energy Policy* 39(6), 3832–3844.

The ESD in Sri Lanka, likewise, incorporated national and local banks, micro-credit agencies, the Ceylon Electricity Board, national ministries, NGOs such as Practical Action, and industry trade groups such as the Grid-connected Small Powers Developers Association. Polycentricity ensured that a "diversity of responses" facilitated the efficient performance and self-organization of a program in the face of "unanticipated conditions."[17]

Each of our six successful programs, moreover, shared costs over at least three broad sources of funding: the government, the private sector (including banks and financing institutions), and communities themselves. GS receives support from the national government's Infrastructure Development Corporation Limited, a state-owned bank, its customers, and the company itself, as it funnels any profit back into research and operations. The REDP in China, REP in Laos, and REAP in Mongolia relied on a combination of support from international development banks, national governments, and communities. The REDP in China specifically gave market development funds but only if solar energy companies were willing to cover at least half of the costs with their own money. From 1996 to 2005, the REDP in Nepal received about one-quarter of the project's funding from the government, one-quarter from households and communities, and one-half from external donors. Consequently, national government, local banks, microhydro functional groups, village development committees, and communities all had a stake in each project. The ESD in Sri Lanka benefited from cost-sharing among development donors, financial institutions, vendors, and end-users—tailoring financial requirements for different stakeholders and project components. In other words, a demand-driven approach was adopted to design multiple financial models to suit multiple stakeholder needs. Each of the six case studies received a mix of loans and grants from international donors such as the World Bank and ADB, rather than purely "free money."

The REDP in Nepal was specially designed to improve accountability and hedge against corruption. It was formally institutionalized in Village Development Committees, the lowest level of governance in Nepal, and also Microhydro Functional Groups, working committees that must meet at least once a month to maintain and manage each plant. As one expert we spoke with explained, "Nepal has had no election at the local scale in 15 years, so no accountability and transparency exists at that level. The REDP explicitly acknowledged this, and avoided local politicians intentionally by relying on these two task forces to implement microhydro projects."

In the case of India, a multilayered service delivery model was present in each project, whereby the Project Implementing Agency formed a Village Energy Committee consisting of representatives of villagers and the local government.

17   Nair, S. 2010. Evolution of Approaches towards Evaluation of Adaptive Responses, Presentation to the Evaluating Adaptive Responses to *Climate Change Workshop of the South Asia Conference of Evaluators ("The Evaluation Conclave,")*, The Lalit Hotel, New Delhi, India, October 25.

However, consultation among all stakeholders including the MNRE, state governments, VECs, and *panchayats* was poor, leading to delays and frequent disagreements.

In Indonesia, the BPPT as the principle implementing agency did very little to adjust the project design to properly involve financial institutions, SHS vendors, and end-users in the project—mainly letting the Asian Financial Crisis take its toll. As a result, at the end of the project, two out of the four participating banks had gone bust, only one out of the six approved SHS dealers remained in business, and a disappointing 8,054 out of 200,000 sales were achieved.

Similarly, in Malaysia, SREP was supposed to be managed by the Malaysian Energy Commission as the implementing agency. However, the Commission did not have the authority to enforce SREP, reconcile complaints, or expedite projects. With limited oversight, it was always at odds with the utility companies and/or participants when trying to pursue the interests of the project. As such, there was very little multi-stakeholder participation.

In Papua New Guinea, the TSLP was designed to be implemented by the Department of Education in six provinces. For this purpose, only one bank (the TS&L) was chosen, which customers were reluctant to patronize and, moreover, did not have sufficient outlets in the target areas. SHS suppliers were selected based on "low cost" rather than the "ability to make and distribute high quality SHS units" and therefore proved incompetent. In the end, there was not a functioning chain of command and a lack of interaction, understanding, and trust between the key stakeholders, resulting in the collapse of the project.

Now, to be explicit: involving large numbers of actors in an energy development project is not a panacea per se; part of this lesson necessitates choosing *strong* or *strategic* partners, something elaborated on below when discussing "political support" and "project champions." One practitioner in Sri Lanka lamented that "for every good NGO doing work in the region, 5 to 6 bad ones exist—not all are equal—and involving too many can create a real catfight rather than coordination. The key is prioritizing and involving only the best."

Affordability

All successful case studies offered financial assistance to overcome the first cost hurdle related to renewable energy technology; in some cases this took the form of financing from microcredit, in other cases it took the form of an ESCO model. The emphasis, in other words, was on affordability rather than installed capacity. Effective programs also saw programmatic costs decline over time, sometimes due to the phased reduction of incentives (such as grants or subsidies), sometimes improvement in management as experience is gained, sometimes due to lower equipment prices from economies of scale, sometimes due to competition, sometimes a combination of all four.

GS's entire mission continues to revolve around providing microcredit to rural homes so they can purchase SHS, biogas digesters, and ICS. China's REDP investigated numerous mechanisms that would encourage consumer credit access for purchasing SHS, including consumer banks, rural credit cooperatives, and the PV companies themselves, though ultimately it did not favor a large credit facility. It was, however, highly competitive in selecting participants—only the strongest companies were invited. The REAP in Mongolia gave low-interest loans, credits, and rebates to reduce the cost of SHS and small-scale wind turbines, and it also relied on bulk purchases of technology to keep costs low. Both the REDP in Nepal and ESD in Sri Lanka overcame the first cost hurdle through grants and the availability of low-interest loans to consumers. The REP in Laos overcame financing issues with its ESCO approach.

The more successful programs also saw costs in technology, development, and management decline over time. GS has seen the price of solar modules and components drop significantly as more are designed and manufactured domestically. The REDP in China saw the price of SHS drop 24 to 35 percent over the lifetime of the program, depending on system size and location. In Nepal, the REDP's costs plummeted from more than $16,000 per installed microhydro kW (including installation and capacity development costs) in 1996 to less than $6,000 per installed kW in 2006.

In India, even though test projects were supported by 90 percent grants from the government, in most cases revenue management was virtually absent. Consumers were reluctant to pay for services due to the persistent down time plaguing systems, and poor revenue flows further diminished interest in maintaining system operations, creating a vicious cycle. Moreover, disagreements and confusion over fuel contracts only added to the expense, and therefore reduced the affordability of VESP projects.

The credit component of the Indonesia SHS Project failed to adjust to the economic realities of the Asian Financial Crisis. With little domestic manufacturing and assembly, the program was entirely dependent on foreign components, which meant the regional crisis saw the devaluation of Indonesian currency and a threefold increase in the costs of SHS systems. Moreover, dealers decided to export systems out of Indonesia to take advantage of the strong US dollar, meaning the program in essence subsidized overseas investment rather than the domestic expansion of energy access.

In Malaysia, insufficient tariffs that were not based on sound economic principles and cost recovery hobbled participants that had to go through the arduous process of applying for Clean Development Mechanism (CDM) credits to make a slim profit. In addition, Malaysian banks were unfamiliar with renewable energy projects and were therefore reluctant to lend to the sector.

In Papua New Guinea, the financial institution chosen to provide credit was one that customers did not visit and which did not have the necessary outlets in the targeted areas. Another substantial impediment was the complete lack of financial literacy on the part of the end-users. As one respondent put it, there is "no sense of savings, credit, debt, or even money." Given these circumstances, it is not difficult to imagine why a renewable energy project using a market-based approach like the TSLP gained little traction in the country.

Capacity Building

Successful programs all undertook some degree of capacity building. Variants of this lesson include strengthening the technical or managerial capacity of domestic firms and institutions; outsourcing to international consultants when capacity is lacking; awarding research grants to manufacturers; improving the business practices of participating organizations; and emphasizing commercial viability (and the ultimate goal of self-sustaining local markets for renewable energy).[18]

Institution building can take a variety of forms. The REDP in China and REAP in Mongolia, for instance, focused on strengthening the ability for SHS equipment manufacturers to improve designs and performance through research grants; the ESD and REDP in Nepal did the same to microhydro designers and manufacturers. The REDP in China also established a verification and claims tracking database so inspectors could track technical performance across the solar industry. The REDP in Nepal elected village leaders to learn how to manage microhydro dams and awarded grants for electricity connections to schools, hospitals and the replanting of trees used in microhydro construction. The REDP spent more on developing capacity and training than on technology. One study calculated that the total cost of installing microhydro systems under REDP was $14.3 million spread over 1996 to 2006, yet 56 percent was spent on capacity development and institutional strengthening, only 44 percent was spent on hardware such as electromechanical machinery, civil works, transportation, and turbines.[19] In Laos, the REP enabled planners to develop a GIS database to coordinate electrification efforts, and it actually outsourced project components for SHS distribution off-grid to the French company, IED.

The REDP in China, ESD in Sri Lanka, REDP in Nepal, and REAP in Mongolia also directed funds to improve the logistics and management of participating institutions. These funds enabled companies and government ministries (when applicable) to build their own capacity in accounting, auditing, sales, and promotion, including tips on revenue collection and accountability as well as advertising campaigns and market analysis.

The necessity of such capacity building was confirmed by a University of Berkeley California study which noted that the best energy development programs directed resources not at particular technologies or projects, but to institutions

---

18 Natalia Magradze, Alan Miller, and Heather Simpson, *Selling Solar: Lessons from More Than a Decade of Experience* (Washington, DC: Global Environment Facility/ International Finance Corporation, 2007).

19 Clemens et al. 2010.

themselves so that they, in turn, could distribute technology and implement projects in perpetuity.[20]

Additionally, successful projects tend to strongly emphasize commercial viability, and aim to create market conditions where their own programs are no longer needed. Put another way, effective partnerships build local capacity so that a self-sustaining renewable energy market can function without external support or dependence on international actors. As one consultant with experience implementing renewable energy projects in dozens of countries commented, "energy services must always be paid for, at a fair cost ... Once you give something away for free, you better be prepared to give it away for free forever."

This lesson about not giving technologies away has been confirmed by one wide-ranging survey of renewable energy in developing countries, noting that donations without cost recovery can actually destroy markets. [21] Many state-financed renewable energy projects following the "give technology away for free" model resulted in damaged technologies, since people tend not to take care of things they do not have to pay for. The study also found that such approaches can inhibit commercial markets as consumers come to expect more donor aid and will wait rather than pay market prices. It lastly found that subsidies are unlikely to lead to sustainable markets unless they explicitly create conditions whereby they are no longer needed, and that they can undermine private investments.

For our case studies, successful projects either resulted in a self-sufficient local market or in a follow-up, larger project. As one example, GS must break even or operate at a profit in Bangladesh or it loses money and goes bankrupt. This is why households must make down payments on technologies and then pay off microfinance loans so that they own them, ensuring that sales of technology are autonomous and self-sustaining. "We don't count on goodwill alone to diffuse technology," one GS employee told us, "we're a company and we collect payments." The REDP in China prompted 28 solar energy companies to develop and expand their rural business enterprises, creating what the World Bank has described as "commercially sustainable sales and service networks that reach the deepest rural areas in China and increasingly they are selling in world markets."[22] Similarly, the REAP in Mongolia has improved the quality and reduced the unit costs of SHSs, nurturing the development around 50 private companies distributing PV equipment in the country. In Nepal and Sri Lanka, only microhydro projects that would earn returns on community investment greater than about ten percent were eligible for support.

---

20   Kammen, D.M. 1999. Bringing Power to the People: Promoting Appropriate Energy Technologies in the Developing World. *Environment* 41(5),10–15, 34–39.

21   Eric Martinot, Akanksha Chaurey, Debra Lew, Jose R. Moreira, and Njeri Wamukonya, "Renewable Energy Markets in Developing Countries," *Annual Review of Energy and the Environment* 27 (2002), pp. 309–348.

22   World Bank 2009. *Implementation, Completion and Results Report* (IBRD-44880 TF-22642).

In terms of follow-ups, the ESD in Sri Lanka was responsible for the establishment of "more than 200 organizations with over 2,000 people commercially involved in grid-connected, off-grid community and household based renewable energy systems". Its successor projects, namely RERED and RERED-AF, built upon these achievements and expanded to include more technologies and components such as income-generating activities. In 2011, Laos embarked on REP II, a follow up project to further increase rural electrification achieved in the REP I, through SHSs and mini-grids powered by solar, pico/microhydro and wind; and further improve the financial performance of EdL. Nepal's Renewable Energy for Rural Livelihood (RERL) was also initiated in 2011 upon the successful completion of the REDP with a focus on increasing energy access and enhancing rural livelihoods especially for poor women and socially excluded groups.

Certainly, none of the four failed case studies discussed in this book meaningfully built capacity, created a self-sufficient market, or resulted in follow-up projects. Only about half of the biogas systems installed under the VESP in India are still operational, even though the project ended in 2011. Out of the six SHS dealers approved in the Indonesia project, only one remains, and participating banks had little to no rural presence, with two of the four closed down during the Asian Financial Crisis and the other two barred by the Bank of Indonesia from giving credit until much later in the project. The Indonesian program spent 95 percent of its budget on technology and only five percent on capacity building. In Malaysia, the SREP allocated no funds for capacity building, and TNB viewed the development of the renewable energy market as detrimental to its profit margins and thus vigorously opposed projects. In Papua New Guinea, selecting SHS vendors based on a least cost approach resulted in approved bids being too low. One firm went bankrupt, whereas the other lacked strong networks in rural areas and provinces it was supposed to serve.

Flexibility

A recognition that programs will need to be flexible in the technologies they include is a common element among our successes. Though not all case studies validate this lesson, an appreciation that not everything can be planned, a realization that programs must be altered to account for changing circumstances, has proven to be an important factor in overcoming implementation challenges.

As briefly indicated above in the section on Appropriate Technology, all of our successes were diverse in the eligible technologies they included, customizing particular technologies to different sets of end-users. GS in Bangladesh did not presume what customers wanted, it asked them and then involved them in its programs. In a way each program targets those with different energy service needs, and eligible technologies vary by price. Those with livestock, straw, dung, corn, agricultural waste, and poultry can subscribe to biogas, those in need of light or television, a SHS, those using prodigious amounts of fuelwood, an ICS. Within each program, participants can select fiberglass or brick biogas units ranging from smaller household capacities to larger community-scale systems, SHS ranging from 10 Wp to 130 Wp (or even configured as micro-grids), and variously sized cookstoves. As one community leader told us:

> Everyone in Bangladesh can benefit from at least one GS program, the ultrapoor with no land and cows can still afford an ICS, most poor households can afford a moderately sized SHS, the top 15 percent of the poor, the upper part, can go for biogas, in this way GS is very versatile, it's programs are matched seamlessly to end user needs, each at a different rung of the energy ladder.

The REDP in China and REP in Laos featured SHSs of various sizes, the REAP in Mongolia did the same and also promoted variously sized wind turbines, the REDP in Nepal ranged widely in the types of microhydro units it supported, and the ESD in Sri Lanka distributed different types of SHS and microhydro dams. The REDP in China was also adjusted its disbursement of TI and MSDF funds, initially awarding $1.50 per Wp for every SHS passing certification standards but increasing it to $2 per Wp to offset higher compliance costs.

In addition, being strategic about whether energy access will be provided by national electricity grids, micro-grids, or off-grid technologies featured in four of our cases. Nepal's REDP, Sri Lanka's ESD, and the REP I in Laos all met this criterion, with separate components tailoring grid, micro-grid, and off-grid solutions to local circumstances. The ESD went so far as to specifically target different beneficiaries: SHS for rural households, grid-connected MHPs for village cooperatives and NGOs, grid-connected MHPs for tea estate management companies and IPPs. The REAP in Mongolia, similarly, pursued a two-pronged strategy of isolated units for some herders but micro-grids for others living near *soum* centers.

By contrast, all of our failures exhibited a degree of rigidity in selecting technology: only technologies fueled by biomass were eligible under the VESP in India; only one type of SHS was eligible in both the Indonesian and Papua New Guinean cases; and the SREP excluded wind energy and promoted no solar energy systems because it was predisposed towards landfill gas capture and waste from the palm oil industry.

Some successful (and unsuccessful) cases adjusted targets and extended deadlines as well. The underlying element to this criterion is possessing adaptability to local events and unforeseen circumstances. China's REDP ended up cancelling their finance component for SHS midway through because targets were being met without it, whereas Indonesia's SHS canceled theirs because of ostensible lack of interest. The REP in Laos revised their grid-connection targets midway through the program and increased solar service rates to account for the climbing price of system components and of the kerosene lamps they were competing with. The REDP in Nepal extended its deadlines twice, the ESD extended its deadline once, to account for ongoing civil wars in each country. The ESD in Sri Lanka was also revised midterm to include microfinance institutions as PCIs to overcome the difficulties facing its SHS component. The VESP in India was cancelled *early* for not achieving its targets, the Indonesian SHS program revised its targets downward to account for a regional financial crisis and switched its subsidies to be per installed capacity rather than number of systems sold. The SREP in Malaysia and TSLP in Papua New Guinea extended their original deadlines as well, though this was because targets were *not* being met, rather than the other way around.

The implicit lesson here is that flexibility in program design and management can assist in expanding access and meeting goals.

Evaluation and Monitoring

Relying on feedback and independent monitoring to adjust programs or partnerships as needed has proven instrumental in our successes. This can be broken down into having independent evaluations of project performance, and/or penalties for noncompliance.

In terms of monitoring and evaluation, it is important to first have clear goals and targets to monitor. All of our six successes, and even our four failures, explicitly set out the goals they wanted to accomplish, whether it was the empowerment of women solar technicians (GS), distribution of 50,000 SHS and WTS in Mongolia (REAP), the construction of 250 kW of village hydro systems in Sri Lanka (ESD), or the installation of 500 MW of qualified renewable electricity capacity in Malaysia (SREP). Only the best programs however, selected strong participating organizations and rigorously evaluated performance.

For example, participating organizations in GS in Bangladesh are independently peer-reviewed and evaluated every year on their performance. As one manager told us, "getting good quality implementing agencies is the hardest part, especially here in Bangladesh, which is why every single one is chosen carefully through peer review and competitive bidding." GS also conducts frequent surveys and evaluations to ensure their program targets are being met. World Bank evaluations of REDP appeared in three separate products: the Implementation Completion and Results (ICR) Report, the ICR Review, and the Project Performance Assessment Report. Similar evaluations were performed by the independent parts of the World Bank (e.g., teams that had no stake or direct involvement in the program), independent researchers from the UNDP, and/or independent consultants for the REP in Laos, REAP in Mongolia, REDP in Nepal, and ESD in Sri Lanka—often before, during, and after the projects commenced. REAP conducted surveys measuring consumer acceptance before and after purchases of equipment.

Three of our four failures—India, Indonesia, and Papua New Guinea—also had external evaluations completed. Such evaluations are useful for not only informing managers if they are accomplishing their goals, but also for increasing interaction and feedback among project designers, end users, and manufacturers.[23]

Five of our successes had penalties for poor performance or non-compliance with programmatic standards, as alluded to above. GS in Bangladesh closed down offices failing to meet expectations and can refuse to pay participating organizations if they do not meet financial standards set by the national government. The REDP in China fined multiple manufacturers millions of dollars for failing to meet technical standards, as did the REAP in Mongolia. The UNDP and World Bank required that only microhydro schemes in Nepal with optimal financial rates of return, and high capacity factors, would continue to receive support under the scheme. The ESD in Sri Lanka relied on a balance of flexibility and various check-and-balance mechanisms to keep projects from becoming derailed. For example, villagers interested in the village hydro scheme were required to establish an electrical consumer society, solicit a project developer, fulfill various administrative requirements, and obtain verification from a chartered engineer before they could go ahead with loan application and building the system. The Federation of Electricity Consumer Societies banned poorly performing manufacturers and a few VECs from participating in the ESD. And financial rates of return had to be above 20 percent for eligible SHS projects and above 24 percent for microhydro projects. Only in Laos were PESCOs able to accrue debts, or distribute substandard technology, without penalty—creating a serious obstacle that continues to afflict the program today. Congruently, our four failures had little information about users and credit, and inconsistent, short-lived, or nonexistent monitoring.

Political Support

Effective programs received consistent political support, including a dedicated or experienced implementing agency, integration with other policies and regulations, and a clear project champion.

A key role for the government in any development project is to provide political leadership and facilitate an appropriate institutional and regulatory environment. During their initial phases, projects can benefit greatly from favorable legal intervention, the integration or harmonization of policies, the creation or alteration of existing institutions, and the development well-targeted incentives. GS benefitted from a national effort directed by IDCOL to distribute SHSs; the REDP in China benefitted from no less than six other renewable energy programs operating concurrently sponsored by a half-dozen international donors; the REP in Laos was supplemented by other "phases" of electrification. The REAP in Mongolia has benefitted from a national renewable portfolio standard and feed-in tariff, the

23   Barnes, Douglas F. and Willem M. Floor. 1996. "Rural Energy in Developing Countries: A Challenge for Economic Development." *Annual Review of Energy and Environment* 21, pp. 497–530.

REDP in Nepal saw ancillary support from related projects including bilateral programs operated by USAID, GTZ, and Danida in addition to a World Bank sponsored Nepal Power Development Project. The ESD in Sri Lanka was backed by public subsidies for electrification as well as a renewable portfolio standard and tax credits; its advocates also changed the Sri Lankan constitution so that off-grid and grid-connected developers had the legal right to distribute and sell electricity. At the same time, rapid political turnovers, such as eight administrations in less than ten years in Bangladesh, and ongoing civil wars in Nepal and Sri Lanka meant that programs effectively *avoided* politicians as much as possible.

Conversely, our four failures did not see widespread political support, or saw it evolve in the wrong direction. In India, local political leaders often gave fuelwood away for free, mitigating the incentive to pay for a biogas system. In Indonesia, national subsidies for kerosene and grid extension plans functionally eroded the desirability of purchasing SHS units. In Malaysia, SREP was a cornerstone of the country's Fifth Fuel Diversification Plan and featured prominently in subsequent development plans, but lacked support from state and local policymakers, and it had to constantly swim upstream against subsidies for fossil fuels. Even worse, in the case of Papua New Guinea, the TLSP was an extension of the failed Solar Lighting for Rural Schools (SLRS) program funded by a Japanese donor, without any genuine commitment from the private sector, teachers unions, or school boards.

Designating a proper coordinating or regulatory agency with a strong mandate and well-defined responsibilities can provide the oversight and coordination needed for a project to succeed. In fact, an autonomous, experienced, or strong implementing agency can be essential in ensuring that committed staff implement the project in a competent manner. As a nonprofit company, GS is entirely under the control of its managers. The REDP in China had a tough but fair PMO that oversaw its subcomponents, monitored progress, managed difficulties, and levied fines against participants that failed to meet standards. As the CEO of one Chinese solar company recently put it, "who wins the clean energy race really depends on how much support the government gives."[24] The REP in Laos, REAP in Mongolia, and ESD in Sri Lanka all had dedicated project management units or offices established by the World Bank and UNDP. The REDP in Nepal was implemented under the direction of the Alternative Energy Promotion Center.

Indeed, it may be that a project's champions—those who advocate strongly for its cause—can help determine successful or failure. A champion can be a key leader in the community, a renowned political figure, an industry group, a grassroots organization, or even political pressure from a vulnerable community who publicly air their grievances or aspirations. In this regard, GS had the benefit of famous leaders including the Nobel Laureate Muhammad Yunus, the founder of the Grameen Bank, and a strong founding director Dipal C. Barua. The close personal connections between Yunus and President Bill Clinton played a role in

---

24   Bradsher, Keith. 2010. "China Builds Lead on Clean Energy." *International Herald Tribune* September 9, p. 1, 16.

USAID giving GS a $4 million grant in the late 1990s, so that the organization was able to raise funds and grow at critical moments of its existence.

In a similar vein, the ESD in Sri Lanka had a strong AU behind it plus a collection of banks and trade and industry associations. These included the Solar Industries Association, Grid Connected Small Power Developers Association, Small Hydropower Developers Association, and Federation of Electricity Consumer Societies in addition to a dedicated NGO (Practical Action) and impassioned microfinance institution (SEEDS). DFCC Bank, the AU, was even willing to "initially sustain losses from operations for the first three years." As one respondent explained, "one of the reasons the ESD succeeded is because the government stayed out—the private sector just ran with it, guided by the AU." The REDP in Nepal was likewise supported by grassroots microhydro functional groups; the REP in Laos was favored by a collection of local and provincial organizations.

In contrast, our four failure case studies suffered from lack of meaningful support. The VESP in India had little consistency between projects. The implementation of the Indonesia SHS Project waned from a lack of coordination amongst the government agencies involved. Although BPPT had all the trappings of a proper implementing agency, in actual fact it was only interested in the technological development component of the project with no real desire to coordinate and oversee other subcomponents. Relatedly, Malaysia's SREP found real opposition from TNB, the national utility, and neither SCORE nor the Malaysian Energy Commission had any real political power to counter TNB's dissatisfaction. The TSLP in Papua New Guinea was plagued by litigation in the courts, and an open disagreement between the World Bank and PNG Sustainable Energy regarding the direction the program should be taking.

Chapter 14

Conclusion

For roughly half of the global population, existence—and energy consumption—is remarkably distinct from the lifestyles most people in industrialized countries have become accustomed to. Imagine a daily ritual without consistently hot showers or baths, no indoor lighting at night, poorly cooked food, and debilitating health problems associated with indoor air pollution. Think about life with no steady pumping of water for drinking and irrigation, few televisions, mobile phones, or computers, and limited access to the fruits of modern civilization.

For those in the developing world, the search for energy fuels and services is an arduous, consistent, exhausting battle. Women and children spend hours each day carrying fuel and water loads often in excess of their weight, time they could otherwise utilize on productive work or education, with calamitous consequences on their health, their natural environment, and their community.

Yet many if not most developing countries still lack the capacity and technology to shift to more sustainable and affordable supplies of energy without external assistance. One survey of the 24 least developed countries in the world found that 22 of them each had less than 1 percent of their region's total energy resources.[1] With scarce energy reserves of their own, these countries must rely either on the global trading system or development assistance from benevolent middle and upper income countries, both outside of their control.[2]

Expanding renewable energy access for rural and increasingly poor communities, nonetheless, is a daunting task. Those without electricity or dependent on traditional fuels tend to have income levels, purchasing power, and consumption levels far below what private companies and electric utilities typically deem profitable, reluctance further attenuated by the inaccessibility of these communities to national electricity grids. Public officials, like their private counterparts, prioritize investments in urban infrastructure where most of their constituents reside, and they often subsidize grid electricity to existing customers instead of expanding access to rural ones or incorporating off-grid technologies. Some international donors continue to focus only on pushing particular

---

1   United Nations Economic and Social Commission for Asia and the Pacific [UNESCAP]. *Energy Security and Sustainable Development in Asia and the Pacific.* (Geneva: UNESCAP, ST/ESCAP/2494, April 2008), p. 185.

2   John Ruggie, *Interim Report of the Special Representative of the Secretary-General on the Issue of Human Rights and Transnational Corporations and Other Business Enterprises* (Geneva: U.N. Doc E/CN.4/2006/97).

technologies instead of holistically utilizing energy services to improve standards of living and productivity.

Six of our case studies presented in this book, however, have overcome many of these difficulties and have shown to rapidly expand energy access and meet millennium development goals simultaneously. Four have succumbed to their challenges and failed to accomplish their targets.

For example, Grameen Shakti in Bangladesh partnered a government-owned infrastructure bank with a nonprofit organization and international donors. It demonstrates that giving communities and women access to credit and training can facilitate gender empowerment and accelerate the deployment of SHSs, biogas units, and ICSs.

China's REDP partnered national and provincial governments with the World Bank and local renewable energy manufacturers. It shows that stringent technical standards, including fines for manufacturers producing inferior equipment, coupled with phased grants to enhance the capacity of renewable energy companies, can create a self-sustaining market for SHSs.

The REP in Laos, a partnership between the national electric utility, an international consultancy, the World Bank, rural companies, and village leaders, proves that expanding the grid and broadening access to off-grid SHS do not have to tradeoff with each other. Instead, subsidized electricity rates for more affluent consumers pay for the extension and expansion of the national grid in addition to the distribution of off-grid solar energy equipment.

The REAP in Mongolia maximized private sector participation and established strong technical and regulatory standards related to the dissemination of SHS and small-scale WTS to nomadic herders. It demonstrates the necessity and effectiveness of providing market development support such as financing for advertising campaigns, after-sales service, and warranties as well as tailoring programs to local needs. For example, local banks were not keen to provide credit to herders as they are nomads and cannot be traced most of the year, something the REAP addressed directly.

The REDP in Nepal weaved together a government ministry, the World Bank, the UNDP, and community leaders. It disbursed some of the revenues from microhydro electricity generation directly to local villages so that they could put energy towards more productive uses such as irrigation, carpentry workshops, bakeries, food preparation, and farming.

Sri Lanka's ESD exhibits the strength of having a well-designed financial model and credit facility, a dedicated implementing agency, and stakeholder participation and capacity building. These elements have garnished innovations in technical and social development that continue to drive a thriving renewable energy market throughout the country.

These six successes contrast with our four unsuccessful case studies.

The VESP in India demonstrates that political will at the provincial and local levels is key to project success. There, a policy framework and clear vision for all stakeholders, including clear lines of authority, was lacking. The VESP implies that such obstacles can be overcome by evolving user-friendly products, strengthening supply chains, and establishing after-sales service mechanisms.

The SHS Project in Indonesia suffered from an inexperienced World Bank promoting its own ideas on how to shift from investing in larger, centralized power projects to smaller scale SHS projects, an improperly designed financial model that calculated risks and incentives poorly, limited capacity building for local stakeholders, and minimal effort to inquire about what Indonesian end-users desired or needed—as they were merely seen as passive energy consumers. The Indonesian government was almost totally uncoordinated in the implementation of the project. Private sector players and financial institutions lacked knowledge about solar energy and were risk averse, and many users remained uninformed about SHS or uninterested in electricity altogether. In essence, the Indonesia SHS Project failed because it did not adapt or adjust to local circumstances and needs.

In Malaysia, the SREP agonized from design faults including capacity caps, a lengthy approval process, lack of monitoring, exclusion of stakeholders, and few pre-feasibility studies. Fragmentation and lack of cohesion with other Malaysian energy policies, notably continued subsidies for natural gas and oil as well as conflicts with state guidelines and policies, further weakened the program. Rather than promote the SREP, the state-owned utility TNB vehemently opposed it and used a variety of tactics, such as interconnection fees, costly feasibility studies, and delays to discourage projects. Electricity tariffs under SREP did not match true production costs, were not based on sound economics, and did not provide cost recovery for project developers.

The TSLP in Papua New Guinea installed only a single SHS, despite expending millions of dollars. A difficulty distributing high quality panels and components in PNG and a lack of product diversity constrained consumer choices. Limited financial literacy, the upfront cost of purchasing a system, weak participating financial institutions and suppliers, and cultural norms and low public awareness further blunted attempts by the program to distribute SHS to teachers.

Despite the uniqueness of these cases, however, this chapter does offer six salient conclusions for those interested in eradicating energy poverty throughout Asia and beyond.

The Socio-technical and Polycentric Nature of Renewable Energy Adoption

First, each of our case studies confirms the socio-technical nature of the acceptance, or rejection, of new technologies. This means that program managers need not only focus on making high quality, standardized technology that works well, they must also get the price signals and financing right, mold cultural values and expectations, spread awareness, align political regulations, and build institutional capacity.

This finding has radical implications for energy development assistance programs, which by and large continue to emphasize technical research, certification, and lowering prices as their most salient priorities when transferring renewable energy equipment to developing countries; politics and culture often takes what one respondent called "a back seat."

Furthermore, our case studies convincingly confirm that successful renewable energy programs focus simultaneously on building the capacity of companies and firms alongside government institutions and the communities that intend to use renewable energy equipment, essentially operating at multiple geographic and institutional scales.

For example, successful programs emphasized building the capacity of the private sector through grants for basic research, grants or loans for manufacturing and production, and efforts to standardize, certify, and test technology. Programs also trained staff at rural electricity companies, cooperatives, and manufacturers in things like tariff setting, metering, billing, revenue management, accounting, and auditing, as well as the formulation of business plans, and advertising and marketing campaigns. Other activities supporting the private sector included staff recruitment and education, the establishment of rural outlets, and the expansion of product inventory.

Concomitantly, capacity in the public sector was strengthened in our successes through the ability to recruit new staff, devise electrification and renewable energy deployment master plans, operate databases and new computer systems, and conduct feasibility studies and resource assessments. In addition, assistance was given to enhance the ability to monitor and evaluate projects, report results, arrange bulk purchases, and create or upgrade government research and testing laboratories.

In tandem with efforts building private and public capacity, successful programs lastly enhanced the capacity of communities and end-users. This included supporting the distribution of sales catalogs, displaying sample systems, and transmitting reliable information to potential customers through targeted television, radio, and newspaper marketing campaigns. It involved the establishment of warranties that protected consumers and the expansion of centers that offered after-sales service and maintenance. It sometimes necessitated hosting workshops and conferences, sponsoring door-to-door promotional efforts, demonstrations, and road shows. One innovative aspect of community capacity building involves funds which mandate that part of project revenue becomes invested directly back into the community, either for maintenance, or to cover community training related to the use of income generation and the operation of end-use machinery. Another innovative aspect is the use of community mobilization funds where project revenues get cycled into grants for schools, health posts, clinics and hospitals.

Essentially, the lesson here is that successful projects are not only socio-technical; they address multiple parts of the supply chain, multiple parts of the project's management, crisscrossing "supply" and "demand" for energy and electricity as well as the private, public, and household sectors, making them polycentric.

The Commonality of Challenges

Second, given this socio-technical nature of renewable energy adoption, many of the challenges facing programs and projects—even successful ones—fall into similar categories. They transcend technical, economic, political, and socio-cultural dimensions, but adhere to a common format, summarized in Table 14.1. What is interesting about this table is that some challenges vex even the most successful programs.

Moreover, the five most frequently identified challenges are providing energy services to the very poor (mentioned in six case studies), difficulty attaining credit and financing (six case studies), difficulty maintaining profitability (six case studies), constrained institutional capacity (eight case studies), and fragmented energy decision-making and authority (six case studies). All five of these challenges reside predominately in either the economic or political and institutional domains, implying that such factors may be more influential than those in the technical and socio-cultural domains.

The Role of Culture

Third, although we believe that many of our conclusions transcend any particular country, some cases exist where well-designed programs with sufficient financial incentives promoting high quality technology have nonetheless failed to convince households to adopt renewable energy technology due to cultural reasons. Moreover, as technologies get rolled out to more and more remote areas, they invariably come into contact with more isolated local cultures.

For example, in Bangladesh, an aversion to pigs has prevented predominately Muslim households from adopting biogas units that would run on pig waste, despite the fact that such waste is much more efficient than dung. Other households refuse to purchase cookstoves at all because they are uncomfortable with the idea of piping in gas from livestock and human excrement, which they see as "impure."

In Nepal, a social norm against collecting revenue for electricity inhibits the profitability of some microhydro schemes. Some believe hydroelectric facilities should serve the community for free, and that poor families should not have to pay for electricity. As one village leader told us, "almost everyone in rural Nepal is poor, so a cultural stigma exists against charging rural households tariffs for microhydro electricity that match costs."

In Papua New Guinea, SHSs have been prone to unusually high rates of vandalism, sabotage, and theft. Under a *wantok* system rooted in tribal traditions, clans there share resources. Solar panels, which benefit a particular house or individual instead of the community, assault this system of *wantok*. Tribal communities have therefore smashed hundreds of solar panels or, worse, threatened their owners. One village elder told us that he would "never" want to

Table 14.1 Common challenges facing our case studies

	Bangladesh	China	Laos	Mongolia	Nepal	Sri Lanka	India	Indonesia	Malaysia	Papua New Guinea
Technical										
Dependence on imported materials and manufacturing constraints				X	X			X	X	X
Technology development and standardization	X								X	
Limited eligible technologies								X	X	X
Weak after-sales service and maintenance		X			X			X	X	X
Poor operational performance (including low load factors or adoption rates)					X	X	X	X	X	X
Logistical challenges (including installation or project delays)		X					X	X	X	X
Inconsistent fuel supply							X	X	X	
Economic and financial										
High upfront costs				X				X	X	X
Providing affordable services to the very poor	X				X	X	X	X	X	X
Difficulty attracting financing		X			X		X	X	X	X
Difficulty collecting revenue or ensuring profitability			X				X	X	X	X
Negative impact on local markets (including bankruptcies and/or market saturation)	X	X	X			X	X	X	X	X

	Challenge							
Political and institutional	Constrained institutional capacity	X	X	X	X	X	X	X
	Commitment to fossil fuels	X		X	X	X	X	X
	Commitment to grid-electrification	X	X	X	X	X	X	X
	Rapid changes in government or project tenure	X					X	
	Fragmentation in energy decision-making	X		X	X	X	X	X
	Aid dependency		X			X		X
Socio-cultural	Low consumer awareness	X	X	X	X	X	X	X
	Unrealistic expectations and lack of familiarity	X			X		X	
	Community disagreement or opposition		X	X		X		X

Note: An 'X' signifies that a program meets a particular challenge.

purchase a solar panel because if "if I did put one in my village, but not all of the surrounding villages, they would kill me."

We have also met individuals with odd expectations about what renewable energy technologies were capable of, or what they needed to function. In Papua New Guinea a school principal believed a single small solar home system could produce enough energy to power a computer lab, a copy machine, lights in every classroom, and a range of appliances, when in fact it could only light four lamps continuously. In Bangladesh, one family thought they needed to dismount their solar home system and take it "for a walk" every day so it "wouldn't get tired." In Nepal one Buddhist mother thought she needed to cover her solar home system with leaves to make it "part of nature," another thought theirs would best "dry laundry" and placed socks and underwear on the solar array shown in Figure 14.1—behavior that likely continues to this day since it "worked" at drying the laundry, even though that system would be producing electricity far less efficiently. Yet another community leader in Nepal said they "intensely disliked" hydroelectricity because it "wasn't Marxist" (whatever that means). In Mongolia, one family thought their solar system would work indoors and had positioned it inside their *ger* as a coffee table. In Sri Lanka, some homeowners thought that a SHS would catch their roof on fire, or require an entirely new type of roof, which was untrue. Another family thought they had to "add soap" to their waste to "make it clean" before it would work in a biogas digester.

Figure 14.1   A family misusing solar equipment near Lukla, Nepal

In short, these examples reveal that cultural attitudes and social expectations can play as significant a role as failed price signals, poorly designed programs, and improperly aligned regulations in impeding the use of off-grid renewable energy applications. They also imply that no matter how well developed or perfected a given renewable energy technology or energy system becomes, it could have little to no impact without systematic and scientific efforts to ensure such technologies are culturally compatible.

Culture, moreover, need not always serve as an impediment; it can catalyze programmatic success. In Bangladesh, a taboo against letting men inside homes during the day enabled GS to empower women as entrepreneurs and technicians, and in Sri Lanka, a culture of *shramadana* convinced communities to give their own time or materials for the civil works and construction of microhydro units. These examples, and others, imply that culture can both impede or accelerate renewable energy adoption.

The Salience of Failure

Our forth conclusion concerns failure. One thing that sets this book apart from others is its simultaneous focus on successes and failures. Admittedly, the notion of each is difficult to pin down, since the terms are charged and politicized. As John F. Kennedy famously proclaimed, "victory has a thousand fathers, but defeat is an orphan." We took a rather simple notion of failure to mean that a "successful" project met its goals or produced benefits that exceeded costs; a "failed" project did not meet its goals or had costs that outweighed benefits.

But researching our four "failures" was admittedly difficult. We were essentially investigating what did *not* exist: systems that had never been installed, units that had never been sold, sites that were especially hard to access, programs that people wanted to forget. We were also investigating what people preferred we *not* assess, and our path was made all the more difficult by denied meetings and difficulties getting reliable data.

What this process—and the striking lessons from our four failures—suggests to us is that both the energy policy community and perhaps the development community as a whole need to better understand the dynamics of failure alongside the better known reasons for success. One respondent from the World Bank even mentioned that "in all likelihood, there are more failures than successes—you just never hear about them. It is easier to fail than to succeed, but all too often failures are swept under the carpet." An appreciation of past experiences and the barriers encountered from such failures can accelerate feedback and lead to new programs that actually meet some or all of their targets.

But this lesson goes deeper. As one respondent from our PNG case study criticized:

It annoys me to see organizations put positive spin on projects that more rightly deserve to be described as mediocre or failures. Sure, we can't always have wins, but too often outcomes are smoothed over with soft rhetoric meaning the reasons behind failed projects don't make it far enough back up the food chain to make a difference to how future projects are designed and implemented ... I suppose learning about PNG culture and politics, and past failures, might require more effort than TSLP managers were willing to provide.

There is nothing wrong with idealism and taking comfort in programmatic successes, but examples like that of our four unsuccessful case studies should motivate energy and development professionals to remain critical and self-reflective. Practitioners should strive to a more nuanced and culturally appreciative view of the role that renewable energy plays in economic development, rather than viewing a particular technology or pathway as an automatic precursor to improved standards of living.

Our case studies further imply that energy access is a higher development goal, not a lower one, and as such, attainment of energy security and reductions in energy poverty will happen only when more basic needs, such as the repayment of debt, financing of education, and satisfaction of community responsibilities are accomplished.

Recommendations for Best Practices

Fifth, though Chapter 11 offers an almost staggering list of recommendations and lessons for policymakers, scholars, researchers, analysts, bilateral donors, and multilateral institutions wishing to implement renewable energy projects abroad, we would like to reemphasize a few best practices here.

Some of the most important, to us, include:

- Do not rely predominately on western consultants, foreign manufacturers, or expertise and knowledge from "outside" of a country. Instead, start by communicating directly with those that intend to use a particular energy service or technology from the "inside."
- Do not dictate overly rigid targets and goals, or restrict eligible technologies, and allow for a degree of flexibility and decentralization in implementing projects.
- Be clear in the targets to be achieved, and consistently monitor and meaningfully evaluate progress towards them.
- Prioritize the affordability of energy services above and beyond the desired installed capacity of a specific technology.
- Actively promote community participation and ownership, do not "give away" systems for free or over-subsidize programs to the extent they hurt existing markets and firms.

- Do not take consumer awareness or information about renewable energy for granted, and instead direct resources at marketing and promotion.
- By the same token, do not assume that the energy services needs and expectations of consumers will remain constant. Renewable energy programs should more readily anticipate changes or technological upgrades as part of a longer-term energy services delivery plan.
- Strongly emphasize after-sales service and maintenance, ensuring that rural populations or technicians care for technologies.
- Couple energy services with income generation and employment.
- Distribute responsibilities among different institutional partners, involving a diversity of important stakeholders, especially those in the private sector and civil society, including companies, nongovernmental organizations, cooperatives, consumer groups, and trade associations.
- Offer financial assistance to overcome the first cost hurdle related to purchasing technology; in some cases offer financing from microcredit, or self-help groups, or cooperatives, or even direct tariffs to consumers.
- Aim for a self-sustaining renewable energy market which can function without external support or dependence on international actors; in essence, move from projects to self-sustaining markets.

Particular experiences will always differ according to culture and context, yet when designed appropriately, programs that adhere to these lessons can quickly and effectively accelerate the adoption of renewable energy technologies in the areas of the world that most urgently need them.

Moreover, many of these lessons have been "around" for decades in the literature on program design and development studies, but appear to have been forgotten. It might be that existing practitioners are too busy "doing" energy development to "think" and learn from each other, or read the academic literature on the topic, including books such as this one. It could be that those that learn some of these lessons, due to rapid turnover within an organization, or sheer frustration, leave the field altogether, resulting in a sort of institutional amnesia. It could be that some development practitioners are biased, interested only in pushing a particular, preselected technology that they or their institution can personally benefit from; or they may suffer from hubris, believing they truly know what's best without the need for external advice. Regardless of the case, many of these lessons urgently need relearned, and they do suggest the need for a new way of thinking about energy and development.

An Emerging Energy Development Paradigm

Lastly, and in line with this goal of crafting a new way of thinking, our case studies subtly, but surely, underscore a somewhat revolutionary approach to the entire concept of energy and development.

Table 14.2   Three paradigms of renewable energy access

	Donor Gift Paradigm (1970s and 1980s)	Market Creation Paradigm (1990s and 2000s)	Sustainable Program Paradigm (2010s)
Actors	One, usually a government or just one development donor	Multiple government agencies and/or multilateral donors	Multiple public, private, and community stakeholders
Primary Goal	Technology diffusion	Market and economic viability	Environmental and social sustainability
Focus	Equipment, often single systems	Multiple fuels (e.g. "electricity" or "fuelwood")	Energy services, income generation, institutional and social needs and solutions
Standardization	Little standardized between projects	Some standardization	Standardized with certificates, testing regimes, and national standards
Implementation	One time disbursement	Project evaluation at beginning and end	Continuous evaluation and monitoring
After-sales Service and Maintenance	Limited	Moderate	Extensive
Ownership	Given away	Sold to consumers	Cost-sharing and in-kind community contributions
Awareness Raising	Technical demonstrations	Demonstrations of business models	Demonstrations of business, financing, institutional, and social models

As Table 14.2 shows, the "classic" approach to energy development assistance in the 1970s and 1980s focused exclusively on single fuels, implemented by a central agency, involving a single financer or borrower. Such projects predominately favored large systems, with little local participation, and strong subsidies for fuels or capital equipment. They overemphasized energy production and installed capacity, and focused on universal access, regardless of the level of development.[3] This approach essentially believed that developed countries should "give" technology and assistance away to developing countries out of a sense of moral obligation.

A new paradigm arose in the 1990s and 2000s, emphasizing the need for multiple fuels, a market approach supported by technical assistance and training, and multiple borrowers pushing smaller, high-return projects. Greater involvement with other donors was incentivized, and projects became more orientated toward consumer demands, integration with broader development efforts, and higher levels of local involvement and investment. This approach essentially assumed that if prices of technology could be brought down beyond a certain point, or local manufacturing established up to a particular threshold, the adoption of renewable energy would become self-sustaining.[4] Or, as one consultant put it, "companies mistakenly believed that if they provided a great product, or delivered a great service, customers would purchase it—nothing else was needed."

The newest paradigm, espoused by many of our case studies, maintains a focus on polycentrism, or the involvement of multiple actors from multiple spheres. Programs extend beyond technological diffusion and market viability to encompass goals such as environmental sustainability, the reduction of greenhouse gas emissions, and local job creation. Their focus is on energy services and income generation rather than fuels or equipment, but they still recognize the necessity of high quality, standardized, and certified technology. Evaluation and monitoring are continuous, after sales service and maintenance are extensive, and communities share costs and in-kind contributions to projects. The paradigm recognizes that the definition of energy affordability varies according to market segments, relative to incomes, market applications, and geography, and that broader social and political factors must be promoted alongside technology and market development.

---

3 Drawn in part from Barnes, D.F. and Floor, W.M. 1996. Rural Energy in Developing Countries: A Challenge for Economic Development. *Annual Review of Energy and Environment* 21, 497–530; Martinot, E. 2001. Renewable Energy Investment by the World Bank. *Energy Policy* 29, 689–699; and Martinot, E. et al. 2002. Renewable Energy Markets in Developing Countries. *Annual Review of Energy and the Environment* 27, 309–348.

4 Magradze, N., Miller, A., and Simpson, H. 2007. *Selling Solar: Lessons from More Than a Decade of Experience*. Washington, DC: Global Environment Facility/International Finance Corporation.

The Need for Further Research

To complement these six conclusions, we see at least seven areas of future inquiry that would help advance our understanding of energy poverty and renewable energy access.

First and foremost, given the hundreds and perhaps thousands of renewable energy access programs around the world, our sample size of 10 cases is remarkably small. Expanding the analysis beyond our case studies, within and outside of Asia, to confirm or disprove our factors would be expedient.

Second, a more formalized, empirical, quantitative assessment of the influence that each of our 42 factors in Chapter 13 has on renewable energy access would be insightful. Especially worthwhile would be conducting factor analysis, fuzzy analysis, and problem trees of the distinct variables responsible for successes and failures.

One study, for example, relied on factor analysis and noted that in India, failure to adopt or maintain microhydro units relates strongly to variables such as unequal income distribution within villages, disputes over water or water shortages, and changes in management. Poorly trained operators, natural calamities like floods and landslides, misuse or inefficient monitoring and little to no funds for maintenance also explain non-adoption.[5] Another study conducted a factor analysis of the conditions explaining successful microhydro operation in Sri Lanka, and it noted the benefits of having a strong village leaders, the assistance of the local provincial council, a stream with strong flow and head to obtain consistent power output, an "easy" site for construction, and community agreement concerning cost sharing and electricity distribution.[6]

Similarly, fuzzy analysis could assess whether energy poverty programs exhibit robust correlations with high per capita incomes, the achievement of millennium development goals, a greater number of patents related to innovations in renewable energy technology, less corruption, lower fuel prices, and a variety of other desirable attributes. The depiction of "problem trees" can also help visualize how a collection of seemingly disparate barriers all relate to a few central causes— such as declining SHS adoption in Sri Lanka being correlated with grid extension.[7]

Other research could explore which factors matter the most, which should be "weighted" more heavily, or as more influential than others. Our data implies, somewhat tenuously, that economic and institutional obstacles resonate more strongly than technical and socio-cultural obstacles, though all dimensions are meaningful.

---

5   Ete, M. and Prochaska, F. 2009. Determinants of Success and Failure of Community Based Micro Hydro Projects,. Proceedings of the *Energy and Climate Change in Cold Regions of Asia Seminar*, April 21–24, Ladakh, India, 33–35.

6   Deheragoda, C.K.M. 2009. "Renewable energy development in Sri Lanka: With special reference to small hydropower," *Tech Monitor* (November/December), pp. 49–55.

7   Enno Heijndermans, *Sri Lanka: Renewable Energy for Rural Economic Development (RERED) Project PV Component Trouble Shooting Workshop Report* (Colombo: Sri Lanka, November 17–18, 2009).

Third, research distinguishing project inputs—how projects ought to be designed and implemented, what goes "in" to them—from outputs—what those projects accomplished, what society got "out" of them—would be valuable, as well as the sequential or temporal nature of some factors. Some of our 42 factors in Chapter 13, such as community involvement in design, product guarantees, and research funding, are clearly inputs. Others, such as enhanced access, lowered fuel consumption, and improved technology, are outputs. Some factors, such as political support, can be both an input (creates momentum to get a project started) or an output (a project that results in national legislation or leadership). Same with resources: having strong resource potential can be an input motivating the creation of a program, or an output of a successful project (such as the successful implementation of an energy plantation or woodlot).

In terms of timing, it could be that some of our factors, such as feasibility studies and technical standards, are prerequisites for others, such as capacity building and reliability, meaning that a certain sequence of factors is needed to achieve success. One Sri Lankan respondent argued that the ESD worked so well merely because "it came at the right window, it was all about timing." We functionally mix these inputs and outputs together, as well as their timing, in Chapter 13, but demarcating them would be helpful.

Fourth, it is conceivable that when undertaking the case studies, we overlooked other drivers that strongly correlate to project successes and failures. For example, it may be possible that there is a correlation between the pace at which electricity demand increases and support for a particular renewable energy program. Geography and resource potential certainly play strong roles, with Nepal and Sri Lanka adopting microhydro units because they are well suited to the fast moving, mountainous rivers throughout each country; Indonesia and Papua New Guinea are well adapted to solar energy in part because of their archipelagic nature. Population density, national population, and education can play a convincing part: part of the reason Grameen Shakti has grown so quickly is because it has a captive market of more than 160 million potential customers in Bangladesh (the seventh largest population on the globe), China is the most populous country in the world, contrasting sharply with the relatively small number of people in Malaysia or Papua New Guinea (though this argument does not explain the tribulations of the VESP in India, the second most populous country in the world).

Moreover, the high literacy rate of rural villagers in Sri Lanka contributed to the progress of the ESD while the lower literacy rates of those in Indonesia and Papua New Guinea inhibited dissemination. The global financial crisis and increasing consensus on the need for international action on climate change also influenced our case studies, whether by motivating political leaders to initiate projects or by affecting the price of materials and commodities related to renewable energy systems, especially those for international carbon credits. Research exploring which internal or external factors were unnoticed in our analysis would be valuable.

Fifth, and perhaps most challenging, is the issue of causality. We relied on a rudimentary notion of causality that presumed a case study accomplished its results if communities started adopting renewable energy systems while that program was in operation. In reality, a variety of other factors beyond the program itself could have influenced adoption rates. The REDP in Nepal, for example, was operating in conjunction with more than six other development projects, including the World Bank-led Nepal Power Development Project; the REDP in China ran almost simultaneously with the NEDO PV Program, Silk Road Brightness Program, and the Township Electrification Program, among others. Research determining causality in such complex policy environments and energy marketplaces would be exceedingly beneficial.

Sixth, developing more sophisticated methods of determining the costs and benefits of a particular project would be wise. Displaced emissions, jobs, and cost benefit analysis do a reasonable job identifying successes and failures, but these tools cannot quantify everything, and looking at what was cost beneficial—did it save more than it cost—is inherently different than looking at what was cost effective—was it the cheapest option compared to all alternatives. In Sri Lanka particularly, some respondents suggested that the SHS subcomponent was cost beneficial but not cost effective, and that ESD funds should have instead been spent on cookstoves or allocated to the off-grid VHP scheme.

Seventh, we have explored three primary energy services and affiliated technologies in this book—renewable sources of electricity for lighting and telecommunications, improved cookstoves for eating and heating, and microhydro units for mechanical power—but are missing at least a fourth: mobility. Although precise numbers are difficult to obtain, as it is less studied, a significant proportion of the world population has transportation choices constrained by lack of infrastructure, fuel scarcity, the distances or time involved with travel, expense, or a combination of them all.

Put another way, the energy poor often need more energy, and pay more for it, to go efficiently or quickly where they need to go, or they merely forego "going" altogether. For both the rural and urban poor, low mobility, regardless of the technology or mode of transport involved, stifles the attainment of better living standards. It reduces the ability to earn income, strains economic resources, and limits access to education and health services and markets. Moreover, lack of mobility can impact other dimensions of energy poverty. It can influence the availability or price of kerosene (lighting), fuelwood collection times (impacting heating and cooking), and transporting processed products to market (mechanical power).

If we are right about the complex multidimensionality of energy poverty, then we functionally need to increase the topics and metrics we associate with it—starting with mobility, expanding to include other neglected energy services—and we need to be more sensitive to those that suffer from it. Perhaps only then will we be able to effectively lift the world's poorest out of their persistent state of energy deprivation.[8]

---

8   Sovacool, BK, C Cooper, M Bazilian, K Johnson, D Zoppo, S Clarke, J Eidsness, M Crafton, T Velumail, and HA Raza. "What Moves and Works: Broadening the Consideration of Energy Poverty," *Energy Policy* 42 (March, 2012), pp. 715–719.

Conclusion

Despite these caveats and suggestions for future research, we do have enough knowledge from our case studies to conclude that renewable energy diffusion can be effective at meeting national and programmatic targets for electrification and access, sometimes ahead of schedule and below cost.

We know that the inclusion of multiple stakeholders in program design, implementation, and evaluation can enhance the efficacy of renewable energy deployment. The involvement of women's groups, multilateral donors, rural cooperatives, local government, manufacturers, nongovernmental organizations and other members of civil society, and even consumers, can increase both the performance and legitimacy of partnerships. They improve performance since input from multiple stakeholders can accelerate feedback; they improve legitimacy since programs with a broader base of support, and community involvement, are less likely to be opposed, protested, or even attacked physically during civil wars and internal conflicts.

We know that the most effective way to expand access to renewable energy through partnerships necessitates a shift in how most development practitioners conceive of energy technology and program structure.[9] Effective partnerships emphasize markets and energy services for customers, rather than technologies. They go beyond merely equipment supply to assess income generation, applications, and user foci. They consider the economic viability of renewable energy technology as only one piece of the puzzle alongside policy formation, financing, institutional capacity, and social and cultural needs. They usually require national or local champions, either in the form of institutions or individuals. Successful programs share not only the rewards of building sustainable renewable energy markets, but also the risks.

Lastly, and perhaps most importantly, we know that investments in renewable energy bring benefits that far exceed their costs. In some cases these include improvements to household income and standards of living, in others productivity and community development. In others they bring technological reliability and quality, and reductions in cost. In still others they encompass significantly reduced greenhouse gas emissions and rates of deforestation. Investments in renewable energy technologies and programs represent one of those rare cases where not only households and small enterprises benefit, but also companies, regulators, and society at large.

---

9   This paragraph draws significantly from Martinot et al. 2002, p. 311.

Appendix
List of Institutions Interviewed/Visited

Agency for the Development and Implementation of Technology (BPPT) Indonesia
Agricultural Development Bank of Nepal Limited (ADBL)
Alpha Solar Systems Sri Lanka
Alpha Thermal Solar Systems Sri Lanka
Alstom Hydro
Appropriate Technology Projects Papua New Guinea
Asian Development Bank
Bangladesh Agricultural Research Initiative
Bangladesh Centre for Advanced Studies
Bangladesh Institute of Development Studies
Bangladesh Ministry of Environment and Forestry
Bangladesh Rural Electrification Board
Bangladesh Solar Energy Society
Bangladesh University of Engineering & Technology
Bank Rakyat Indonesia (BRI)
Barefoot Power Systems
Beijing Jike New Energy Technology Development Company
Bell Palm Industries Sdn. Bhd. Malaysia
Borneo Resources Institute Malaysia
Bright Green Energy Foundation
Butwal Power Company Limited
Centre for Environment, Technology, and Development Malaysia
Ceylon Electricity Board
Chinese Academy of Sciences
Chinese Ministry of Environmental Protection
Chinese Ministry of Finance
Chinese Ministry of Land and Resources
Chinese Ministry of Science and Technology
Chinese Renewable Energy Society Photovoltaic Committee
Chinese Solar Home System Dealiers
CIMB Niaga Bank
Clean Energy Nepal
Conservation International
Consultancy and Professional Services Sri Lanka
Consultants for Resource Evaluation
Daesung Group Mongolia

Dawa Indonesia
DFCC Bank Sri Lanka
Eco Power Group of Companies Sri Lanka
Eco-Ideal Consulting Sdn. Bhd. Malaysia
Electricité du Laos
Energy Authority of Mongolia
Friends of the Earth
German Technical Cooperation (GTZ)
Gesang Solar
Global Environment Facility
Grameen Bank
Grameen Shakti
Grameen Technology Center
Hashakee Power (Pvt.) Ltd
Hatton National Bank
Hilful Fuzul Samaj Kallyan Sangstha
HK Mart Mongolia
Hydrosolutions Nepal
Indian Ministry of New and Renewable Energy
Indian Ministry of Power
Indian State Electricity Boards
Indonesia Ministry of Energy and Mineral Resources (MEMR)
Indonesia Ministry of Finance (MENKEU)
Indonesia Ministry of Research and Technology (MENRISTEK)
Indonesia National Development Planning Agency (BAPPENAS)
Indonesia State Electricity Company (PLN)
Indonesian Institute for Energy Economics
Indonesian Institute of Sciences (LIPI)
Indonesian Renewable Energy Society
Infrastructure Development Company Limited Bangladesh
Institute for Participatory Interaction in Development Sri Lanka
Institute of Strategic & International Studies Malaysia
Integration Consulting Group Mongolia
International Center for Integrated Mountain Development (ICIMOD)
International Finance Corporation
Japan International Cooperation Agency
JD Energy Systems
Kathmandu University
Kub-Berjaya Enviro Sdn. Bhd. Malaysia
Langkawi Development Authority
Lanka ORIX Leasing Company Sri Lanka
Laos Institute for Renewable Energy
Laos Ministry of Energy and Mines
Laos Provincial Electrification Service Companies

Laos Village Off-grid Promotion Scheme
Laos Water Resources and Environment Administration
Light up the World Foundation
Macquarie University
Malaysia Board of National Economic Advisory Council
Malaysia Corridor Development Unit, Prime Minister's Department
Malaysia Economic Planning Unit, Prime Minister's Department
Malaysia Forest Research Institute
Malaysia Ministry of Energy, Green Technology & Water
Malaysia Ministry of Natural Resources and Environment
Malaysia Ministry of Tourism
Malaysia Public Private Partnership Unit, Prime Minister's Department
Malaysia Renewable Energy Research Centre (SIRIM)
Malaysian Energy Commission
Malaysian Palm Oil Board
Mark Marine Services Sri Lanka
McKinsey & Company
Momat Solar Energy
Mon-Energy Consulting Mongolia
Mongolia Energy Association
Mongolia Ministry of Fuel and Energy
Mongolia Ministry of Nature, Environment and Tourism
Nam Theun 2 Power Corporation Laos
National University of Laos
Nepal Alternative Energy Promotion Center
Nepal Electricity Authority
Nepal Hydroelectric and Water Tariff Commission
Nepal Ministry of Energy
Nepal Power Tech Company Limited
Nepal Yentra Shala Energy
Netherlands Development Organization
Nielsen Sri Lanka
OSK Research Malaysia
PACOS Malaysia
Papua New Guinea Department of Petroleum and Energy
Papua New Guinea Institute of National Affairs
Papua New Guinea Power Limited
Papua New Guinea Solar Energy Systems Limited
Papua New Guinea Sustainable Energy Limited
Petronas
Phocos Bangladesh
Practical Action Sri Lanka
PT. Gerbang Multindo Nusantara Indonesia
PT. Mambruk Indonesia

PT. Trimbasolar Indonesia
Qinghai Tianpu Solar Energy Company
Recycle Energy Sdn. Bhd. Malaysia
Renewable Power Sdn. Bhd. Malaysia
Resource Management Associates Sri Lanka
RMA Consulting Sri Lanka
Sabaragmuwa Provincial Government
Sarawak Electricity Supply Company
Sarawak Energy Berhad
Sarawak Government State Planning Unit
Sarawak Hidro Berhad
Sarawak Hidro Sdn Bhd
Sarawak Natural Resources and Environment Board
Sarawak Regional Corridor Development Authority
Sarawak Rivers Board
Sarawak State Government
Sarvodaya Economic Enterprise Development Services
Siemens India
Sikkim Renewable Energy Development Agency
Sime Darby
Sino-Danish Renewable Energy Development Program
SNV
South Asian Association for Regional Cooperation (SAARC)
South Asian Institute of Technology
Sri Lanka Business Development Centre
Sri Lanka Grid-connected Small Powers Developers Association
Sri Lanka Ministry of Finance and Planning
Sri Lanka Ministry of Power and Energy
Sri Lanka Sustainable Energy Authority
Stockholm Environmental Institute
Sunlabob
Tenaga Nasional Berhad Malaysia
The Energy Resources Institute (TERI) India
Transparency International
TÜV SÜD
United Nations Development Programme Bangladesh
United Nations Development Programme Malaysia
United Nations Development Programme Nepal
United Nations Development Programme Renewable Energy Development Program
United Nations Development Programme Sri Lanka
United States Agency for International Development Bangladesh
United States Agency for International Development India
United States Agency for International Development Malaysia

United States Agency for International Development Nepal
United States Agency for International Development Sri Lanka
Universiti Malaysia Sarawak
University of Dhaka
University of New South Wales (Australia)
University of Papua New Guinea
Vallibel Energy Sri Lanka
Vidullanka Plc Indonesia
Westlink Enterprises Limited Papua New Guinea
Whitehouse Institute of Science and Technology
World Bank China
World Bank India
World Bank Indonesia
World Bank Laos
World Bank Mongolia
World Bank Nepal
World Bank Sri Lanka
World Resources Institute
Xining Moonlight Solar Science and Technology Company
Xining New Energy Development Company
Yayasan Bina Usaha Lingkungan Indonesia
Yayasan Pelangi Indonesia

Index